*Climate Policy after*
# KYOTO

# CONTENTS

# Introduction

## *TOR RAGNAR GERHOLM: The debate continues*

At an international conference held in Toronto, Canada, it was stated that the entire planet is threatened by a gigantic environmental catastrophe. "Mankind is conducting an unintended, uncontrolled, globally pervasive experiment whose ultimate consequence could only be second to a global nuclear war. It is imperative to act now."

However, this was ten years ago. Since then the "red alert scenario" has been toned down. Nevertheless, the predictions of the Intergovernmental Panel on Climate Change (IPCC) regarding future climate change and its consequences for human welfare and the natural environment have been sufficiently frightening to give rise to far-reaching political measures that have sought to reduce emissions of carbon dioxide and other greenhouse gases. The most recent expression of this concern was the major conference on climate change held in Kyoto, Japan in December 1997.

The debate on global climate change was not concluded in Kyoto. On the contrary, it will continue to intensify when steps are taken to translate the verbal commitments made at Kyoto into practical politics by means of ratification and legislation in each signatory country, including Sweden.

It would appear that the debate will shift from issues related to climatology, meteorology and science in general to a discussion of the political, economic and social consequences of changes in energy policy that seek to limit and ultimately reduce the use of fossil fuels.

In this book it is our intention to raise some of these issues and provide a forum for experts from areas outside meteorology. Some of these experts have not previously made their voices heard.

We have concentrated on carbon dioxide emissions for several reasons. Firstly carbon dioxide accounts for more than half of the anthropogenic emissions of greenhouse gases. If no measures are undertaken to limit these emissions, their relative share will, according to the IPCC, grow substantially during the next century. Although carbon dioxide will be our primary concern, we are naturally aware that the combined effect of other greenhouse gases may have an effect comparable to that of carbon dioxide.

Carbon dioxide is also of particular importance in view of the direct and apparently inherent link between atmospheric emissions and the use of fossil fuels. Since 80 per cent of our current energy requirements are based on fossil fuels, the relationship between energy use and the environment becomes especially problematical.

Another reason for limiting our discussion to carbon dioxide is that from a Swedish perspective, it accounts for approximately 80 per cent of total emissions of greenhouse gases.

The contributors to this volume will not present their own theories, hypotheses or models. Our analyses will instead be based on IPCC's own reports and on the scientific primary data used by the Intergovernmental Panel on Climate Change. It will become apparent that the conclusions are not as clear cut as they are sometimes assumed to be. In different ways and from a variety of standpoints, several of us have experienced a growing scepticism in relation to the evidence currently available. In our view, there are insufficient grounds for an extensive programme of climate policy measures.

This does not mean that we have adopted a nonchalant attitude to the problem. We are all aware that the earth's climate will be affected if there is a substantial increase in the emissions of carbon dioxide and other greenhouse gases. Hence we recommend measures that will limit and reduce emissions without at the same time generating higher levels of expenditure. Some of us, including myself, are prepared to go further and accept modest increases in expenditure provided that it can be shown with some degree of certainty that these measures will reduce emissions in global terms although not necessarily in Sweden.

In order to provide the reader with a background to the principal issues discussed in this book, we start with a presentation of selected material from the *IPCC Second Assessment Synthesis of Scientific–Technical Information relevant to Interpreting Article 2 of the UN Framework on Climate Change*.

In the introductory essay, *Erik Moberg* seeks in a balanced fashion to examine the underlying special interests that are (potentially) concealed behind the mass media facade surrounding climate policy issues. Among these special interests, the pressure groups associated with science and politics deserve particular attention. The perceptive analyses presented by Moberg and the other contributors provide the reader with critical insights that are not always forthcoming in the debate.

*Frits Böttcher* offers a background to the debate on climate policy and discusses the extent to which scientists should accept both the plaudits and responsibility for allowing climatic questions to take on quite literally global proportions.

*Wibjörn Karlén* shows that contrary to general opinion, the climate of the twentieth century has behaved in a fairly normal fashion. Changes in the surface temperature of the earth of equal magnitude and rapidity as we have experienced during the last one hundred years have occurred during the pre-industrial era as well as apparently during the entire period after the most recent ice age. In a similar vein, in the following essay, *Jarl Ahlbeck* also argues that using conventional statistical methods, there is no evidence to suggest that the greenhouse effect has exerted a significant influence on world climate.

The somewhat self-evident starting point of my own contribution (*Tor Ragnar Gerholm*) is that all predictions regarding future climate changes are based on the assumptions that we make regarding future global energy consumption. These assumptions which relate to circumstances 50 to 100 years from now must be regarded as highly speculative. In another contribution, *Jarl Ahlbeck* presents some interesting results from a statistical analysis that he has recently completed. His study shows that the proportion of carbon dioxide emissions that actually remains in the atmosphere over a longer period – the so-called airborne fraction - declines as the atmospheric concentration of carbon dioxide rises. Accordingly the IPCC's emission scenarios lead to considerably lower estimates for atmospheric $CO_2$ concentrations than those provided by the climate panel. As a result, the projected impact on the climate will be correspondingly lower.

*Richard Lindzen* discusses the shortcomings of existing climate models and their lack of solid empirical foundations. The decisive influence on global climate exerted by the oceans is examined by *Gösta Wallin* who shares Lindzen's view that the IPCC's climate models cannot be used to make predictions about the climatic effects of higher concentrations of greenhouse gases.

*Marian Radetzki* examines the economic analyses of the consequences of climate policy presented by the IPCC's experts. He raises the question whether the substantial levels of expenditure, that will be required to implement climate policy programmes would not be better used to finance more urgently needed reforms, particularly in poorer countries.

The implications of the above argument become abundantly clear in *Karl-Axel Edin's* account of Sweden's own remarkable energy and climate policy that readers in other countries may find to be both instructive and comic.

I hope that the reader will not be put off by the occasional use of mathematical expressions. Their sole purpose is to allow the reader who is able to follow the mathematical reasoning to assess the logic of the argument. At the same time, this will not prevent non-mathematical readers from examining the argument in the accompanying text.

In order to give the reader a chance to draw his own conclusions regarding the validity of our arguments, we have invited *Bert Bolin*, the first chairman of the IPCC, to provide a commentary to the papers presented in this volume. Appropriately, Bolin will be given the final, but not necessarily conclusive, word in the debate.

# IPCC's Basic Documents

# 1.

# IPCC Second Assessment Synthesis – Climate Change 1995

*Sections 2, 3 and 4*

## 2. Anthropogenic Interference with the Climate System

*Interference to the present day*

2.1    In order to understand what constitutes concentrations of greenhouse gases that would prevent dangerous interference with the climate system, it is first necessary to understand current atmospheric concentrations and trends of greenhouse gases, and their consequences (both present and projected) to the climate system.

2.2    The atmospheric concentrations of the greenhouse gases, and among them, carbon dioxide ($CO_2$) methane ($CH_4$) and nitrous oxide ($N_2O$), have grown significantly since pre-industrial times (about 1750 A.D.): $CO_2$ from about 280 to almost 360 ppmv[1], $CH_4$ from 700 to 1720 ppbv and $N_2O$ from about 275 to about 310 ppbv. These trends can be attributed largely to human activities, mostly fossil-fuel use, land-use change and agriculture. Concentrations of other anthropogenic greenhouse gases have also increased. An increase of greenhouse gas concentration leads on average to an additional warming of the atmosphere and the Earth's surface. Many greenhouse gases remain in the atmosphere – and affect climate – for a long time.

2.3    Tropospheric aerosols resulting from combustion of fossil fuels, biomass burning and other sources have led to a negative direct forcing and possibly also

---

[1] ppmv stands for parts per million by volume; ppbv stands for parts per billion (thousand million) by volume. Values quoted are for 1992.

12

to a negative indirect forcing of a similar magnitude. While the negative forcing is focused in particular regions and subcontinental areas, it can have continental to hemispheric scale effects on climate patterns. Locally, the aerosol forcing can be large enough to more than offset the positive forcing due to greenhouse gases. In contrast to the long-lived greenhouse gases, anthropogenic aerosols are very short-lived in the atmosphere and hence their radiative forcing adjusts rapidly to increases or decreases in emissions.

2.4   Global mean surface temperature has increased by between about 0.3 and 0.6°C since the late 19th century, a change that is unlikely to be entirely natural in origin. The balance of evidence, from changes in global mean surface air temperature and from changes in geographical, seasonal and vertical patterns of atmospheric temperature, suggests a discernible human influence on global climate. There are uncertainties in key factors, including the magnitude and patterns of long-term natural variability. Global sea level has risen by between 10 and 25 cm over the past 100 years and much of the rise may be related to the increase in global mean temperature.

2.5   There are in adequate data to determine whether consistent global changes in climate variability or weather extremes have occurred over the 20th century. On regional scales there is clear evidence of changes in some extremes and climate variability indicators. Some of these changes have been toward greater variability, some have been toward lower variability. However, to date it has not been possible to firmly establish a clear connection between these regional changes and human activities.

## Possible consequences of future interference

2.6   In the absence of mitigation policies or significant technological advances that reduce emissions and/or enhance sinks, concentrations of greenhouse gases and aerosols are expected to grow throughout the next century. The IPCC has developed a range of scenarios, IS92a-f, of future greenhouse gas and aerosol precursor emissions based on assumptions concerning population and economic growth, land-use, technological changes, energy availability and fuel mix during the period 1990 to 2100[2]. By the year 2100, carbon dioxide emissions under these scenarios are projected to be in the range of about 6 GtC[3] per year, roughly equal to current emissions, to as much as 36 GtC per year, with the lower end of the

---

[2] See Table 1 in the Summary for Policymakers of IPCC Working Group II.
[3] To convert GtC (gigatonnes of carbon or thousand million tonnes of carbon) to mass of carbon dioxide, multiply GtC by 3.67.

IPCC range assuming low population and economic growth to 2100. Methane emissions are projected to be in the range 540 to 1170 $Tg^4 CH_4$ per year (1990 emissions were about 500 Tg $CH_4$); nitrous oxide emissions are projected to be in the range 14 to 19 Tg N per year (1990 emissions were about 13 Tg N). In all cases, the atmospheric concentrations of greenhouse gases and total radiative forcing continue to increase throughout the simulation period of 1990 to 2100.

2.7    For the mid-range IPCC emission scenario, IS92a, assuming the "best estimate" value of climate sensitivity[5] and including the effects future increases in aerosol concentrations, models project an increase in global mean surface temperature relative to 1990 of about 2°C by 2100. This estimate is approximately onethird lower than the "best estimate" in 1990. This is due primarily to lower emission scenarios (particularly for $CO_2$ and CFCs), the inclusion of the cooling effect of sulphate aerosols, and improvements in the treatment of the carbon cycle. Combining the lowest IPCC emission scenario (IS92c) with a "low" value of climate sensitivity and including the effects of future changes in aerosol concentrations leads to a projected increase of about 1°C by 2100. The corresponding projection for the highest IPCC scenario (IS92e) combined  with a "high" value of climate sensitivity gives a warming of about 3.5°C. In all cases the average rate of warming would probably be greater than any seen in the last 10,000 years, but the actual annual to decadal changes would include considerable natural variability. Regional temperature changes could differ substantially from the global mean value. Because of the thermal inertia of the oceans, only 50-90% of the eventual equilibrium temperature change would have been realized by 2100, even if concentrations of greenhouse gases were stabilized by that time.

2.8     Average sea level is expected to rise as a result of thermal expansion of the oceans and melting of glaciers and ice-sheets. For the IS92a scenario, assuming the "best Estimate" values of climate sensitivity and of ice melt sensitivity and of ice melt sensitivity to warming, and including the effects of future changes in aerosol concentrations, models project an increase in sea level of about 50 cm from the present to 2100. This estimate is approximately 25% lower than the "best estimate" in 1990 due to the lower temperature projection, but also reflecting improvements in the climate and ice melt models. Combining the lowest emission scenario (IS92c) with the "low" climate and ice melt sensitivities and including aerosol effects gives a projected sea-level

---

[4] Tg: teragram is 1012 grams
[5] In IPCC reports, climate sensitivity usually refers to long-term (equilibrium) change in global mean surface temperature following a doubling of atmospheric equivalent $CO_2$ concentration. More generally, it refers to the equilibrium change in surface air temperature following a unit change in radiative forcing (°C/Wm$^{-2}$).

rise of about 15 cm from the present to 2100. The corresponding projection for the highest emission scenario (IS92e) combined with "high" climate and ice-melt sensitivities gives a sea-level rise of about 95 cm from the present to 2100. Sea level would continue to rise at similar rate in future centuries beyond 2100, even if concentrations of greenhouse gases were stabilized by that time, and would continue to do so even beyond the time of stabilization of global mean temperature. Regional sea-level changes may differ from the global mean value owing to land movement and ocean current changes.

2.9   Confidence is higher in the hemispheric-to-continental scale projections of coupled atmosphere-ocean climate models than in the regional projections, where confidence remains low. There is more confidence in temperature projections than hydrological changes.

2.10   All model simulations, whether they were forced with increased concentrations of greenhouse gases and aerosols or with increased concentrations of greenhouse gases alone, show the following features: greater surface warming of the land than of the sea in winter; a maximum surface warming in high northern latitudes in winter, little surface warming over the Arctic in summer; an enhanced global mean hydrological cycle, and increased precipitation an soil moisture in high latitudes in winter. All these changes are associated with identifiable physical mechanisms.

2.11   Warmer temperatures will lead to a more vigorous hydrological cycle; this translates into prospects for more severe droughts and/or floods in some places and less severe droughts and/or floods in other places. Several models indicate an increase in precipitation intensity, suggesting a possibility for more extreme rainfall events. Knowledge is currently insufficient to say whether there will be any changes in the occurrence of geographical distribution of severe storms, e.g., tropical cyclones.

2.12   There are many uncertainties and many factors currently limit our ability to project and detect future climate change. Future unexpected, large and rapid climate system changes (as have occurred in the past) are, by their nature, difficult to predict. This implies that future climate changes may also involve "surprises". In particular, these arise from the non-linear nature of the climate system. When rapidly forced, non-linear systems are especially subject to unexpected behaviour. Progress can be made by investigating non-linear processes and subcomponents of the climatic system. Examples of such non-linear behaviour include rapid circulation changes in the North Atlantic and feedback's associated with terrestrial ecosystem changes.

# 3. Sensitivity and Adaptation of Systems to Climate Change

3.1    This section provides scientific and technical information that can be used, *inter alia*, in evaluating whether the projected range of plausible impacts constitutes "dangerous anthropogenic interference with the climate system", as referred to in Article 2, and in evaluating adaptation options. However, it is not yet possible to link particular impacts with specific atmospheric concentrations of greenhouse gases.

3.2    Human health, terrestrial and aquatic ecological systems, and socio-economic systems (e.g., agriculture, forestry, fisheries and water resources) are all vital to human development and well-being and are all sensitive to both the magnitude and the rate of climate change. Whereas many regions are likely to experience the adverse effects of climate change – some of which are potentially irreversible – some effects of climate change are likely to be beneficial. Hence, different segments of society can expect to confront a variety of changes and the need to adapt to them.

3.3    Human-induced climate change represents an important additional stress, particularly to the many ecological and socio-economic systems already affected by pollution, increasing resource demands, and non-sustainable management practices. The vulnerability of human health and socio-economic systems – and to a lesser extent, ecological systems – depends upon economic circumstances and institutional infrastructure. This implies that systems typically are more vulnerable in developing countries where economic and institutional circumstances are less favourable.

3.4    Although our knowledge has increased significantly during the last decade and qualitative estimates can be developed, quantitative projections of the impacts of climate change on any particular system at any particular location are difficult because regional scale climate change projections are uncertain; our current understanding of many critical processes is limited; systems are subject to multiple climatic and non-climatic stresses, the interactions of which are not always linear or additive; and very few studies have considered dynamic responses to steadily increasing concentrations of greenhouse gases of the consequences of increases beyond a doubling of equivalent atmospheric $CO_2$ concentrations.

3.5    Unambiguous detection of climate-induced changes in most ecological and social systems will prove extremely difficult in the coming decades. This is because of the complexity of these systems, their many non-linear feedback's, and their sensitivity to a large number of climatic and non-climatic factors, all of which are expected to continue to change simultaneously. As future climate extends beyond the boundaries of empirical knowledge (i.e., the documented impacts of climate variation in the past), it becomes more likely that actual outcomes will include surprises and unanticipated rapid changes.

## Sensitivity of systems

### TERRESTRIAL AND AQUATIC ECOSYSTEMS

3.6    Ecosystems contain the Earth's entire reservoir of genetic and species diversity and provide many goods and services including: (i) providing food, fibre, medicines and energy; (ii) processing and storing carbon and other nutrients; (iii) assimilating wastes, purifying water, regulating water runoff, and controlling floods, soil degradation and beach erosion; and (iv) providing opportunities for recreation and tourism. The composition and geographic distribution of many ecosystems (e.g., forests, rangelands, deserts, mountain systems, lakes, wetlands and oceans) will shift as individual species respond to changes in climate; there will likely be reductions in biological diversity and in the goods and services that ecosystems provide society. Some ecological systems may not reach a new equilibrium for several centuries after the climate achieves a new balance. This section illustrates the impact of climate change on a number of selected ecological system.

3.7    **Forests**: Models project that a consequence of possible changes in temperature and water availability under doubled equivalent[6] $CO_2$ equilibrium conditions, a substantial fraction (a global average of one-Third, varying by region from one-seventh to twothirds) of the existing forested area of the world will undergo major changes in broad vegetation types – with the greatest changes occurring in high latitudes and the least in the tropics. Climate change is expected to occur at a rapid rate relative to the speed at which forest species grow, reproduce and re-establish themselves. Therefore, the species composition of forests is likely to change; entire forest types may disappear, while new assemblages of species and hence new ecosystems may be established. Large amounts of carbon could be released into the atmosphere during transition from one forest type to another because the rate at which carbon can be lost during times of high forest mortality is greater than the rate at which it can be gained through growth to maturity.

3.8    **Deserts and desertification**: Deserts are likely to become more extreme – in that, with few exceptions, they are projected to become hotter but

---

[6] See paragraph 4.17 for a description of "equivalent $CO_2$".

not significantly wetter. Temperature increases could be a threat to organisms that exist near their heat tolerance limits. Desertification – land degradation in arid, semi-arid and dry sub-humid areas resulting from various factors, including climatic variations and human activities – is more likely to become irreversible if the environment becomes drier ant the soil becomes further degraded through erosion and compaction.

3.9 **Mountain ecosystems**: The altitudinal distribution of vegetation is projected to shift to higher elevation; some species with climatic ranges limited to mountain tops could become extinct because of disappearance of habitat or reduced migration potential.

3.10 **Aquatic and coastal ecosystems**: In lakes and streams, warming would have the greatest biological effects at high latitudes, where biological productivity would increase, and at the low latitude boundaries of cold- and cool-water species ranges, where extinctions would be greatest. The geographical distribution of wetlands is likely to shift with changes in temperature and precipitation. Coastal systems are economically and ecologically important and are expected to vary widely in their response to changes in climate and sea level. Some coastal ecosystems are particularly at risk, including saltwater marshes, mangrove ecosystems, coastal wetlands, sandy beaches, coral reefs, coral atolls and river deltas. Changes in the se ecosystems would have major negative effects on tourism, freshwater supplies, freshwater supplies, fisheries and biodiversity.

## Hydrology and water resources management

3.11 Models project that between one-third and one-half of existing mountain glacier mass could disappear over the next hundred years. The reduced extent of glaciers and depth of snow cover also would affect the seasonal distribution of river flow and water supply for hydroelectric generation and agriculture. Anticipated hydrological changes and reductions in the areal extent and depth of permafrost could lead to large-scale damage to infrastructure, an additional flux of carbon dioxide into the atmosphere, and changes in processes that contribute to the flux of methane into the atmosphere.

3.12 Climate change will lead to an intensification of the global hydrological cycle and can have major impacts on regional water resources. Changes in the total amount of precipitation and in its frequency and intensity directly affect the magnitude and timing of runoff and the intensity of floods and droughts; however, at present, specific regional effects are uncertain. Relatively small changes in temperature and precipitation, together with the nonlinear effects on evapotranspiration and soil moisture, can result in relatively large changes in runoff, especially in arid and semi-arid regions. The quantity and quality of water supplies already are serious problems today in many regions, including some low-lying coastal areas, deltas and small islands, making countries in these regions particularly vulnerable to any additional reduction in indigenous water supplies.

## AGRICULTURE AND FORESTRY

3.13 Crop yields and changes in productivity due to climate change will vary considerably across regions and among localities, thus changing the patterns of production. Productivity is projected to increase in some areas and decrease in others, especially the tropics and subtropics. Existing studies show that on the whole, global agricultural production could be maintained relative to baseline production in the face of climate change projected under doubled equivalent $CO_2$ equilibrium conditions. This conclusion takes into account the beneficial effects of $CO_2$ fertilization but does not allow for changes in agricultural pests and the possible effects of changing climatic variability. However, focusing on global agricultural production does not address the potentially serious consequences of large differences at local and regional scales, even at mid-latitudes. There may be increased risk of hunger and famine in some locations; many of the world's poorest people – particularly those living in subtropical and tropical areas and dependent on isolated agricultural systems in semi-arid and arid regions – are most at risk of increased hunger. Global woods supplies during the next century may become increasingly inadequate to meet projected consumption due to both climatic and non-climatic factors.

## HUMAN INFRASTRUCTURE

3.14 Climate change clearly will increase the vulnerability of some coastal populations to flooding and erosional land loss. Estimates put about 46 million people per year currently at risk of flooding due to storm surges. In the absence of adaptation measures, and not taking into account anticipated population growth, 50-cm sea-level rise would increase this number to about 92 million; a 1-meter sea-level rise would raise it to about 118 million. Studies using a 1-meter projection show a particular risk for small islands and deltas. This increase is at the top range of IPCC Working Group I estimates for 2100; it should be noted, however, that sea level is actually projected to continue to rise in future centuries beyond 2100. Estimated land losses range from 0.05% in Uruguay, 1.0% for Egypt, 6% for the Netherlands and 17.5% for Bangladesh to about 80% for the Majuro Atoll in the Marshall Islands, given the present state of protection systems. Some small island nations and other countries will confront greater vulnerability because their existing sea and coastal defense systems are less well established. Countries with higher population densities would be more vulnerable. Storm surges and flooding could threaten entire cultures. For these countries, sea-level rise could force internal or international migration of populations.

## HUMAN HEALTH

3.15   Climate change is likely to have wide-ranging and mostly adverse impacts on human health, with significant loss of life. Direct health effects include increases in (predominantly cardio-respiratory) mortality and illness due to an anticipated increase in the intensity and duration of heat waves. Temperature increases in colder regions should result in fewer cold-related deaths. Indirect effects of climate change, which are expected to predominate, include increases in the potential transmission of vector-borne infectious diseases (e.g., malaria, dengue, yellow fever and some viral encephalitis) resulting from extensions of the geographical range and season for vector organisms. Models (that entail necessary simplifying assumptions) project that temperature increases of 3-5°C (compared to the IPCC projection of 1-3.5°C by 2100) could lead to potential increases in malaria incidence (of the order of 50-80 million additional annual cases, relative to an assumed global background total of 500 million cases), primarily in tropical, subtropical and less well-protected temperate-zone populations. Some increases in non-vector-borne infectious diseases – such as salmonellosis, cholera and giardiasis – also could occur as a result of elevated temperatures and increased flooding. Limitations on freshwater supplies and on nutritious food, as well as the aggravation of air pollution, will also have human health consequences.

3.16   Quantifying the projected impacts is difficult because the extent of climate-induced health disorders depends on numerous coexistent and interacting factors that characterize the vulnerability of the particular population, including environmental and socioeconomic circumstances, nutritional and immune status, population density and access to quality health care services. Hence, populations with different levels of natural, technical and social resources would differ in their vulnerability to climate-induced health impacts.

## *Technology and policy options for adaptation*

3.17   Technological advances generally have increased adaptation options for managed systems. Adaptation options for freshwater resources include more efficient management of existing supplies and infrastructure; institutional arrangements to limit future demands/promote conservation; improved monitoring and forecasting systems for floods/droughts; rehabilitation of watersheds, especially in the tropics; and construction of new reservoir capacity. Adaptation options for agriculture – such as changes in types and varieties of crops, improved water-management and irrigation systems, and changes in planting schedules and tillage practices – will be important in limiting negative effects and taking advantage of beneficial changes in climate. Effective coastal-zone management and land-use planning can help direct population shifts away from vulnerable locations such as flood plains, steep hillsides and low-lying coastlines. Adaptive options to reduce health impacts include protective technology (e.g., housing, air conditioning, water purification and vaccination), disaster preparedness and appropriate health care.

3.18   However, many regions of the world currently have limited access to these technologies and appropriate information. For some island nations, the high cost of providing adequate protection would make it essentially infeasible, especially given the limited availability of capital for investment. The efficacy and cost-effective use of adaptation strategies will depend upon the availability of financial resources, technology transfer, and cultural, educational, managerial, institutional, legal and regulatory practices, both domestic and international in scope. Incorporating climate-change concerns into resource-use and development decisions and plans for regularly scheduled investments in infrastructure will facilitate adaptation.

## 4. Analytical Approach to Stabilization of Atmospheric Concentrations of Greenhouse Gases

4.1   Article 2 of the UN Framework Convention on Climate Change refers explicitly to "stabilization of greenhouse gas concentrations". This section provides information on the relative importance of various greenhouse gases to climate forcing and discusses how greenhouse gas emissions might be varied to achieve stabilization at selected atmospheric concentration levels.

4.2   Carbon dioxide, methane and nitrous oxide have natural as well as anthropogenic origins. The anthropogenic emissions of these gases have contributed about 80% of the additional climate forcing due to greenhouse gases since pre-industrial times (i.e., since about 1750 A.D.). The contribution of $CO_2$ is about 60% of this forcing, about four times that from $CH_4$.

4.3   Other greenhouse gases include tropospheric zone (whose chemical precursors include nitrogen oxides, non-methane hydrocarbons and carbon monoxide), halocarbons[7] (including HCFCs and HFCs) and SF6. Tropospheric aerosols and tropospheric ozone are inhomogeneously distributed in time and

21

space and their atmospheric lifetimes are short (days to weeks). Sulphate aerosols are amenable to abatement measures and such measures are presumed in the IPCC scenarios.

4.4    Most emission scenarios indicate that, in the absence of mitigation policies, greenhouse gas emissions will continue to rise during the next century and lead to greenhouse gas concentrations that by the year 2100 are projected to change climate more than that projected for twice the pre-industrial concentrations of carbon dioxide.

## Stabilization of greenhouse gases

4.5    All relevant greenhouse gases need to be considered in addressing stabilization of greenhouse gas concentrations. First, carbon dioxide is considered which, because of its importance and complicated behaviour, needs more detailed consideration than the other greenhouse gases.

## Carbon dioxide

4.6    Carbon dioxide is removed from the atmosphere by a number of processes that operate on different time-scales. It has a relatively long residence time in the climate system — of the order of a century or more. If net global anthropogenic emissions[8] (i.e., anthropogenic sources minus anthropogenic sinks) were maintained at current levels (about 7 GtC/yr including emissions from fossil-fuel combustion, cement production and land-use change), they would lead to nearly constant rate of increase in atmospheric concentrations for at least two centuries, reaching about 500 ppmv (approaching twice the pre-industrial concentration of 280 ppmv) by the end of the 21st century. Carbon cycle models show that immediate stabilization of the concentration of carbon dioxide at its present level could only be achieved through an immediate reduction in its emissions of 50-70% and further reductions thereafter.

4.7    Carbon cycle models have been used to estimate profiles of carbon dioxide emissions for stabilization at various carbon dioxide concentration levels. Such profiles have been generated for an illustrative set of levels: 450, 550, 650, 750 and 1000 ppmv. Among the many possible pathways to reach stabilization, two are illustrated in Figure 1.1 for each of the stabilization levels of 450, 550, 650

---

[7] Most halocarbons, but neither HFCs nor PFCs, are controlled by the Montreal Protocol and its Adjustmens and Amendments.

[8] For the remainder of Section 4, "net global anthropogenic emissions" (i.e., anthropogenic sources minus anthropogenic sinks) will be abbreviated to "emissions".

and 750 ppmv, and one for 1000 ppmv. The steeper the increase in the emissions (hence concentration) in these scenarios, the more quickly is the climate projected to change.

4.8    Any eventual stabilized concentration is governed more by the accumulated anthropogenic carbon dioxide emissions from now until the time of stabilization, than by the way those emissions change over the period. This means that, for a given stabilized concentration value, higher emissions in early decades require lower emissions later on. Cumulative emissions from 1991 to 2100 corresponding to these stabilization levels are shown in Table 1.1, together with the cumulative emissions of carbon dioxide for all of the IPCC IS92 emission scenarios (see Figure 1.2 below and Table 1 in the Summary for Policymakers of IPCC Working Group II for details of these scenarios).

4.9    Figure 1.1 and Table 1.1 are presented to clarify some of the constrains that would be imposed on future carbon dioxide emissions, if stabilization at the concentration levels illustrated were to be achieved. These examples do not represent any form of recomendation about how such stabilization levels might be achieved or the level of stabilization which might be chosen.

4.10    Given cumulative emissions, and IPCC IS92a population and economic scenarios for 1990-2100, global annual average carbon dioxide emissions can be derived for the stabilization scenarios on a per capita or per unit of economic activity basis. If the atmospheric concentration is to remain below 550 ppmv, the future global annual average emissions cannot, during the next century, exceed the current global average and would have to be much lower before and beyond the end of the next century. Global annual average emissions could be higher for stabilization levels of 750 to 1000 ppmv. Nevertheless, even to achieve these latter stabilization levels, the global annual average emissions would need to be less than 50% above current levels on a per capita basis or less than half of current levels per unit of economic activity[9].

4.11[10]    The global average annual per capita emissions of carbon dioxide due to the combustion of fossil fuels is at present about 1.1 tonnes (as carbon). In addition, a net of about 0.2 tonnes per capita are emitted from deforestation and land-use change. The average annual fossil fuel per capita emission in developed and transitional economy countries is about 2.8 tonnes and ranges from

---

[9] China registered its disagreement on the use of carbon dioxide emissions derived on the basis of a per unit economic activity.
[10] The Panel agreed that this paragraph shall not prejudge the current negotiations under the UNFCCC.

1.5 to 5.5 tonnes. The figure for the developing countries is 0.5 tonnes ranging from 0.1 tonnes to, in some few cases, above 2.0 tonnes (all figures are for 1990).

4.12[11] Using World Bank estimates of GDP (gross domestic product) at market exchange rates, the current global annual average emission of energy-related carbon dioxide is about 0.3 tonnes per thousand 1990 US dollars output. In addition, global net emissions from land-use changes are about 0.05 tonnes per thousand US dollars of output. The current average annual energy-related emissions per thousand 1990 US dollars output, evaluated at market exchange rates, is about 0.27 tonnes in developed and transitional economy countries and about 0.41 tonnes in developing countries. Using World Bank estimates of GDP at purchasing power parity exchange rates, the average annual energy-related emissions per thousand 1990 US dollars output is about 0.26 tonnes in developed and transitional economy countries and about 0.16 tonnes in developing countries.[12]

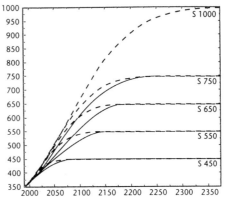

**Figure 1.1 (a).** Carbon dioxide concentration profiles leading to stabilization at 450, 550, 650 and 750 ppmv following the pathways defined in IPCC (1994) (solid curves) and for pathways that allow emissions to follow IS92a until at least the year 2000 (dashed curves). A single profile that stabilizes at a carbon dioxide concentration of 1000 ppmv and follows IS92a emissions until at least the year 2000 has also been defined. Stabilization at concentrations of 450, 650 and 1000 ppmv would lead to equilibrium temperature increases relative to 1990[13] due to carbon dioxide alone (i.e., not including effects of other greenhouse gases (GHGs) and aerosols) of about 1°C (range: 0,5 to 1.5°C), 2°C (range: 1.5 to 4.5°C) and 3.5°C (range 2 to 7°C), respectively. A doubling of the pre-industrial carbon dioxide concentration of 280 ppmv would lead to a concentration of 560 ppmv and doubling of the current concentration of 358 ppmv would lead to a concentration of about 720 ppmv.

---

[11] The Panel agreed that this paragraph shall not prejudge the current negotiations under the UNFCCC.

[12] These calculations of emissions per unit of economic activity do not include emissions from land-use change or adjustments to reflect the informal economy.

[13] These numbers do not take into account the increase in temperature (0.1 to 0.7°C) wich would occur after 1990 because of $CO_2$ emissions prior to 1990.

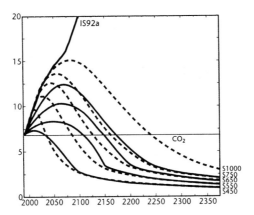

**Figure 1.1 (b).** Carbon dioxide emissions leading to stabilization at concentration 450, 550, 650, 750 and 1000 ppmv following the profiles shown in (a) from a mid-range carbon cycle model. Results from other models could differ from those presented here by up to approximately ± 15%. For comparison, the carbon dioxide emissions for IS92a and current emissions (fine solid line) are also shown.

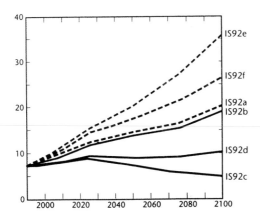

**Figure 1.2.** Annual anthropogenic carbon dioxide emissions under the IS92 emission scenarios (see Table 1 in the Summary for Policymakers of IPCC Working Group II for further details).

25

Table 1.1. Total anthropogenic carbon dioxide emissions accumulated from 1991 to 2100 inclusive (GtC) for the IS92 scenarios (see Table 1 in the Summary for Policymakers of IPCC Working Group II) and for stabilization at various levels of carbon dioxide concentration following the two sets of pathways shown in Figure 1.1 (a). The accumulated emissions leading to stabilization of carbon dioxide concentration were calculated using a mid-range carbon cycle model. Results from other models could be up to approximately 15% higher or lower than those presented here.

| | Accumulated carbon dioxide emissions 1991 to 2100 (GtC)ß | |
|---|---|---|
| **IS92 scenarios** | | |
| c | 770 | |
| d | 980 | |
| b | 1430 | |
| a | 1500 | |
| f | 1830 | |
| e | 2190 | |
| **Stabilization case** | **For profiles A[2]** | **For profiles B[3]** |
| 450 ppmv | 630 | 650 |
| 550 ppmv | 870 | 990 |
| 650 ppmv | 1030 | 1190 |
| 750 ppmv | 1200[4] | 1300[4] |
| 1000 ppmv | – | 1410[4] |

§ For comparison, emissions during the period 1860 to 1994 amounted to about 360 GtC, of which about 240 GtC were due to fossil-fuel use and 120 GtC due to deforestation and land-use change.

* As in IPCC (1994) — see Figure 1.1 (a) (solid curves).

† Profiles that allow emissions to follow IS92a until at least the year 2000 see figure 1.1 (a) (dashed curves).

‡ Concentrations will not stabilize by 2100.

## Methane

4.13 Atmospheric methane concentrations adjust to changes in anthropogenic emissions over period of 9 to 15 years. If the annual methane emissions were immediately reduced by about 30 Tg $CH_4$ (about 8% of current anthropogenic emissions), methane concentrations would remain at today's levels. If methane emissions were to remain constant at their current levels, methane concentrations (1720 ppbv in 1994) would rise to about 1820 ppbv over the next 40 years.

## Nitrous oxide

4.14   Nitrous oxide has a long lifetime (about 120 years). In order for the concentration to be stabilized near current levels (312 ppbv in 1994), anthropogenic sources would need to be reduced immediately by more than 50%. If emissions of nitrous oxide were held constant at current levels, its concentration would rise to about 400 ppbv over several hundred years, which would increase its incremental radiative forcing by a factor of four over its current level.

## Further points on stabilization

4.16   Stabilization of the concentrations of very long-lived gases, such as $SF_6$ or perfluorocarbons, can only be achieved effectively by stopping emissions.

4.16   The importance of the contribution of $CO_2$ to climate forcing, relative to that of the other greenhouse gases, increases with time in all of the IS92 emission scenarios (a to f). For example, in the IS92a scenario, the $CO_2$ contribution increases from the present 60% to about 75% by the year 2100. During the same period, methane and nitrous oxide forcings increase in absolute terms by a factor that ranges between two and tree.

4.17   The combined effect of all green house gases in producing radiative forcing is often expressed in terms of the equivalent concentration of carbon dioxide which would produce the same forcing. Because of the effects of the other greenhouse gases, stabilization at some level of equivalent carbon dioxide concentration implies maintaining carbon dioxide concentration at a lower level.

4.18   The stabilization of greenhouse gas concentrations does not imply that there will be no further climate change. After stabilization is achieved, global mean surface temperature would continue to rise for some centuries and sea level for many centuries.

# Science and Politics

# 2.

# The science and politics of the greenhouse issue

*by Erik Moberg*

The normal or at least currently prevailing view of the significance of the greenhouse effect may be conveniently summarised in terms of a defined threat and a set of specific policy proposals.

**The threat**: As a result of the increased emissions of greenhouse gases induced by human activity, especially carbon dioxide, the earth's atmosphere acts as the glass in a greenhouse. The temperature of the surface of the earth and the adjoining atmospheric layer rises which brings about changes in climate. An important secondary effect is a general rise in the level of the oceans. Together these effects represent a serious threat to substantial parts of humanity.

**The policy proposals**: In order to avert this threat, carbon dioxide emissions must be swiftly and drastically reduced. The most effective way of achieving this goal is to equally swiftly and drastically reduce the use of fossil fuels.

Let us start the discussion by stating that the policy proposals are neither simple nor inexpensive. There is little evidence to the contrary and in fact there would appear to be a wide measure of agreement that these types of measures are exceedingly expensive. Thus the question of the reality of the threat becomes a burning issue. Is the threat that gives rise to the need for these high levels of expenditure a real one? If not, the expenditure would be a gross waste of resources. This line of reasoning points to the risk that expensive environmental programmes will be undertaken on the grounds of an exaggerated threat. However it should also be pointed out that the risk to which attention is normally drawn when demands are made for collective action e.g. in relation to greenhouse gases is the opposite, namely that the measures undertaken will be inadequate in relation to the threat. This risk should therefore also be a subject of discussion.

## The character of the threat scenario

The principal features of the threat scenario, as they have been specified above, are not without controversy. Indeed, there would appear to be widely divergent views on several basic issues and it is important to understand the nature of these disagreements.

Firstly our description of the threat scenario, with the exception of the assertion that the threat is serious, consists entirely of objective statements of fact. It is actually one way or the other. Each individual interpretation of the threat scenario, in its entirety or in part, is either true or false. At the same time, it is important to bear in mind that it is difficult or indeed impossible to resolve these questions of truth in a manner that is equally convincing for everyone. There is for example no simple way of solving these issues by means of controlled experiments. For the conceivable future, any relevant, realistic interpretation must be based on the results of complex models, which are in turn largely dependent on matters of judgement and assessment. The complexity of the models also means that they are inevitably based on the results and preconceptions of several different disciplines in the natural sciences.

These uncertainties regarding the state of knowledge are naturally hardly unique – there are numerous problem complexes of substantial practical importance that have more or less the same features. However, as we already know, science is also full of incontrovertible truths. It is for example possible to predict the movement of heavenly bodies with great precision. Here there is no room for divergent opinions. Of course there may be opponents, but all people knowing anything about the matter will then be in agreement that they are dealing with madcaps. In this context, the dividing line between certain and uncertain knowledge is of great importance. Since uncertainty surrounds the greenhouse threat, the following conclusions may be drawn:

- There is considerable scope for different views even among serious, well-informed researchers.

- Researchers who comment on some aspect of the greenhouse threat that is connected to their own field command special authority.

31

- Researchers who comment on the greenhouse threat in its entirety will need to have a rare intellectual capacity and breadth of knowledge in order to be considered as authoritative spokesmen.

- Interested parties may exploit uncertainty. For instance someone who has an interest in the reality of the threat may advocate this view as long as prevailing uncertainty allows him to do so. On the other hand, an interested party from the other end of the spectrum will tend to adopt a different approach to uncertainty, arguing that the threat is exaggerated or possibly doesn't exist at all.

Finally it should also be emphasised that uncertainty means just what it says. It doesn't mean that one knows that the threat doesn't exist. Hence preventative measures undertaken with safety in mind can be justified on the grounds of uncertainty per se.


## Some important interests

Since interested parties may use the uncertainty surrounding the state of knowledge in relation to global warming to their own advantage, it would seem appropriate to take a closer look at the actual interest groups concerned.

- The oil industry has a natural interest in playing down the threat of global warming since the proposed measures are aimed at their products. Consumers who are heavily dependent on oil may also have a similar interest.

- The same type of interest could also apply to the coal industry and mining employees.

- It is almost equally obvious that interest groups associated with nuclear power would wish to emphasise the reality and seriousness of the threat to the global environment. Nuclear power is a principal alternative to fossil fuel and does not give rise to carbon dioxide emissions.

- Analogous to the case of nuclear interests, current and future producers of biomass i.e. farmers have also a vested interest in emphasising the seriousness of the environmental threat.

- The environmental or green lobby represents a possible interest group that is more difficult to analyse. The difficulty lies in the fact that the views put forward by these organisations may very well be completely sincere and certainly are so in many cases. Many individuals from these circles are convinced that there is a major serious threat to the global environment and that drastic measures are required. This can hardly be called an interest. However there may be others in these groups who have different concerns. They may dislike cars, large-scale enterprises, consumerism, modern industrial society and its life style etc. Here the greenhouse effect is something, which can be emphasised and made use of in order to achieve other goals. In such cases, it could be stated that the threat to the global environment is used as a means for furthering certain interests.

## The roles and interests of states

It is even more difficult to analyse the role of states in this context. Firstly, a state may react in response to the interests of different groups of citizens e.g. by co-ordinating, supporting or suppressing them. However a state may also act on its own account and in some sense may be said to pursue its own special interests. It is perhaps just these interests that are of particular importance in the present context.

A state is seldom reluctant to involve itself in various types of activities. On the contrary, the opposite would appear to be the case. States have a tendency to undertake more and more. An obvious example is public sector expansion. Measured as a share of gross domestic product the public sector, with few exceptions, is continually growing throughout the democratic world. This aspect of the growth of the public sector is well known. Although it is more difficult to establish statistically, the public domain, i.e. the activities undertaken by the state rather than by private individuals, also seems to grow universally and continuously.

The struggle for political power is presumably a principal factor underlying this expansion. In their eagerness to gain the support of the electorate, it has undoubtedly been tempting for politicians and political parties to allocate to themselves new areas of operation rather than just limiting their legislative proposals and promises to their traditional fields. In this way, new virgin territory is opened up to the struggle for political power, which allows politicians to reduce the importance of political conflicts in the traditional arenas. The state thereby takes over certain tasks that have previously fallen within the domain of the private sector. This may occur even though the private provision of these services has not given rise to any problems. Other new areas of state operation may be completely new. This may be either a matter of new demands that have arisen or existing demands that were not met. Some of

33

these new areas of activity may be a reflection of imagined rather than real needs. The expansion of environmental and energy policy in recent decades provides us with numerous examples. The tendency for a state to increase its operations into all kinds of possible and impossible areas thus seems to be irresistible, an inherent part of the state's own dynamic process.

Policies concerned with the greenhouse effect may be seen in this light. In principle, it is perfectly conceivable that the increased emission of greenhouse gases, as has been described above, represents a real threat. If so it is also quite in order that the governments of the world are seeking to do something about this threat to the global environment. The reason, as we shall see below, is that we are dealing with what economists call a public good. However it is also perfectly possible that the threat is exaggerated and that governments give it emphasis in order to further their own interests. The threat may very well be used by ambitious politicians who seek to favour their own causes by expanding the public domain.

It is also interesting to examine this question from the standpoint of energy policy rather than solely concentrate on environmental issues. In contrast to other areas of policy, energy policy is usually profitable for the state. Energy taxation which is a central pillar of energy policy normally yields income receipts that are much larger than expenditure on policy measures. A threat to the environment that provides an argument in favour of this type of policy is therefore an asset. For example, the tax on carbon dioxide emissions that has been justified on the grounds of global warming produces substantial revenue for the Swedish exchequer. Governments in Germany and the United Kingdom may well have been able to refer to the greenhouse threat when closing down unprofitable coal mines.

A government may therefore have substantial interests in emphasising the reality and seriousness of the threat to the global environment. The argument that this threat is exaggerated will consequently be hardly welcome.


## The importance of the constitution

Although the arguments above have a general validity in relation to democratic states, there are also probably differences dependent on the constitution. It is of particular importance here to note that special interests would appear to find it easier to exert their influence in countries that have proportional representation rather than the first-past-the-post system of simple majority voting. The same would appear to hold in parliamentary as opposed to presidential democracies. The first-past-the–post electoral system tends to exclude small parties or single issue parties whereas they flourish under proportional representation. Hence green parties and farmers' parties are to be found in several western European countries that have proportional representation whereas they are absent from countries such as Britain and the United States that have simple majority voting. Furthermore, in contrast to a presidential system, parliamentary democracy usually encourages the growth of unified, disciplined parties that allow interest groups the opportunity to exert direct influence at the government level. Countries that have different types of parliamentary system combined with proportional representation dominate the European Union. Parliamentary government combined with simple majority voting is found in Britain and some Commonwealth countries. The United States has a presidential system and simple majority voting. Consequently the politics of many EU countries are likely to be heavily influenced by special interests. This will also be of substantial importance in relation to dealing with and possibly exploiting the issue of greenhouse gas emission at the government level.

## The researchers' predicament

Political science has given increasing prominence to the idea that politicians are not just governed by considerations of the public good but are also affected by motives of self-interest. A variant of this idea which is somewhat more flattering for politicians states that it is only those politicians who are able to defend their own positions of power that survive as politicians. It could be equally well argued that the same also applies to research and researchers. Naturally the pursuit of truth is a central driving force for many researchers. However this doesn't exclude the possibility that considerable effort may be invested in the pursuit of prestige, status and high income. The acquisition of the latter may be associated with the successful pursuit of truth. However this is not necessarily the case since prestige and status may also be bestowed by others outside the world of academic research.

The central issue in this context is naturally that researchers are largely dependent on the state, which may have substantial interests of its own. In most modern societies, the state is the major financier of research. Since most types of research require substantial funding, it wouldn't be unreasonable to assume that researchers who produce results or work in areas that meet with the approval of the government will have a better chance of survival in the field of research than those who don't. It is a relatively simple matter to find examples of these mechanisms in the social sciences. For instance in Sweden, there is the case of sectoral research in support of the "Swedish model". Similar conditions may also be encountered in those areas of the natural sciences that are subject to complexity and uncertainty and require a considerable element of judgement, as for example in relation to the greenhouse effect. This should be borne in mind when evaluating research into the greenhouse effect.

## The problem of collective action

For a number of different reasons, as we have seen above, the threat to the global environment posed by greenhouse gases may be exaggerated. If so, the measures proposed to deal with this threat are unnecessarily far-reaching. It is, however, also important to examine the problems associated with the counter-measures under the assumption that the threat is real and serious.

Let us for the sake of argument assume that all human beings contribute equally to carbon dioxide emissions and that the damaging effects of these emissions are also spread equally throughout the world. Further let us assume that the benefits from a reduction in carbon dioxide emissions are also evenly distributed among the world's population.

Given these conditions, let us assume that an individual actor or group of actors, e.g. an entire industry, a particular industrial sector, a country or group of countries decides to reduce their carbon dioxide emissions. The total costs of this reduction in carbon dioxide emissions are thus borne by those who undertake the reduction while the world's population in its entirety enjoys the benefits. This example illustrates the complications of collective action. One aspect is that the acting group must be sufficiently big, possibly a considerable proportion of the world's population. Otherwise the group's benefits from the reduction of carbon dioxide emissions would not outweigh its costs, and hence its action would not be rational and likely. Another aspect concerns the strong incentive for "free riding". Here individual groups may try to remain outside a joint programme of emission reductions in order to avoid having to con-tribute towards the costs of the programme while at the same time enjoying its benefits.

Although this mental exercise is schematic and simplified, the resemblance to reality is nevertheless sufficiently close to capture a number of important aspects. If the greenhouse effect poses an actual threat to the environment, it would be accordingly beneficial for nations rather than individuals, since nations are much fewer and bigger than individuals, to take steps to reduce carbon dioxide emis-sions. Even so there are however also a lot of nations and even at the national level it may thus be difficult to reach an agreement about a common plan of action. It is this difficulty that justifies the fear expressed above that the measures undertaken may be inadequate if the threat to the global environment from greenhouse gases is considered to be a matter of real urgency. It takes a long time to achieve results and there are good grounds for starting at an early date. Hence it is perfectly possible to specify a reasonable set of assumptions that would provide a rational basis for present international actions.

## Injuring and injured parties

Let us now instead assume that those who emit carbon dioxide and those who suffer from its consequences belong to largely different groups of the world's population. For instance, it is conceivable that the environmental damage is concentrated in particularly vulnerable areas, especially coastal regions. In this case, we have a relationship between two parties or two groups of parties, an injuring party that emits carbon dioxide and an injured party. This relationship may be examined in both legal and economic terms, and it is also interesting to introduce the possibility of different types of corrective measures. In the previous example, the reduction of carbon dioxide emissions was the only type of corrective measure under consideration.

A complete analysis of the possible relationships between the parties cannot be undertaken here. However the following question may be discussed. Assume that the injured party has the opportunity to free itself from the damage, for example, by paying the injuring parties to undertake the changes in energy production required in order to reduce carbon dioxide emissions. Would it then be rational for the injured party to act in this way? Or, alternatively, would the injured party find it more economically attractive to leave the polluted areas and live elsewhere? If the latter is the case, it would be preferable from a "global household" standpoint to correct the environmental damage by means of population movement rather than energy conversion. In order to see this we can imagine that the injuring party, rather than carrying out the energy conversion, might want to compensate the injured parties for the damage that they have caused by an amount of money equivalent to the costs of energy conversion. Let us furthermore imagine that the injured party, on receiving the financial compensation, decides to move to another area rather than to invest in energy conversion. If so this choice should benefit those who suffered the effects of environmental damage since after having moved, and paid for that, they would have some money left while the consequences for the injuring party would be the same as if it had actually had carried out the energy conversion.

Assuming once again that the threat to the environment from greenhouse gases is real, this mental experiment also captures essential parts of the real situation. There is undoubtedly a need to identify injuring and injured parties, and to analyse the relations between them. There is certainly also a need to closely examine the entire range of corrective measures.

37

## The importance of environmental damage

A real threat to the environment from greenhouse gases would be principally directed towards that part of human activity which is dependent on the soil and the climate i.e. the agricultural sector. When assessing the damage, it is therefore important to note that the agricultural sector's share of total output declines as industrialisation and economic growth expands. Consequently, an increase in costs that affects agriculture as a result of the greenhouse effect will have a more limited impact on the total economy, the lower the agricultural sector's share of total output.

## Conclusions

The threat to the environment from greenhouse gases is considered by many to be both real and serious. Effective corrective measures are therefore seen to be essential. This may be the case and there is also a risk that the logic of collective action will lead to an unfortunate delay in implementing appropriate measures. There are also those who argue that these measures should be undertaken on insurance grounds even though the threat has not been definitely established. If the threat is sufficiently serious and the costs of corrective measures are sufficiently limited, this would be undoubtedly a logical standpoint.

The purpose of this essay has not been to test the validity of these and similar views. Rather, my aim has been to highlight mechanisms which may bring other ideas than the most realistic ones into focus. I have underlined that the greenhouse issue is not only global with possibly fatal consequences but also complex and unresolved scientifically. There is consequently considerable scope for widely varying opinions in even the most serious and learned of academic circles. The question is in this sense still open to scientific debate. This combination of fatefulness and scientific uncertainty provides a fertile ground for interest politics.

It is normally alleged that the oil industry has been successful in minimising the threat to the environment. This may well be the case. There are however interests that operate in the opposite direction such as agricultural interests and the environmental lobby. It is also important to note that governments have their own interests in emphasising the threat to the environment of greenhouse gases. It may be a consequence of the dynamics of political competition or it may simply be a question of government revenue. It is possible that these tendencies are most marked in the type of nations that one finds within the European Union – parliamentary states with proportional electoral systems. Since governments are principal organisers and financiers of research, it is reasonable to assume that research is also affected.

If these assessments are correct, it would also be reasonable to assume that the environmental threat is exaggerated and that substantial resources are being unnecessarily invested to try to avert it.

# 3.

# The use and misuse of science in policy making

*by C.J.F. Böttcher*

## Introduction

Global warming has obtained the status of the most important environmental issue. This remarkable feat came about through the well-orchestrated efforts of an inner circle of science-policy makers within the Intergovernmental Panel on Climate Change (IPCC). They predominated the scientific discussion in order to achieve the necessary "scientific consensus" for politicians to take action. The result is that during the past ten years the topic global warming has been embraced by ministers for the environment, with their civil servants and governmental agencies, by environmental pressure groups and - in their wake - by the media. Within the wheel-work of the UN the striking attention to a subject as complex as climate change could trigger the signing of an international climate treaty after less than five years preparations - a unique event in the annals of international negotiations.

Twenty participants in this intense co-operation between a small group of scientists, politicians and diplomats, wrote an instructive book (1) about it, edited by Mintzer and Leonard of the Stockholm Environment Institute. Their story has the significant title "Negotiating Climate Change: The Inside Story of the Rio Convention". Similar descriptions of the achieved results occur in several chapters of a recent book (2) "International Politics of Climate Change", edited by Fermann, Trondheim University.

The policy making in the case of global warming serves as an instructive example of the use and misuse of science in politics.

# Scientific responsibility and scientific freedom

Scientific responsibility is a delicate subject for several reasons. Non-scientists observe the scientific community with a mixture of some admiration, some anxiety and a lot of misunderstanding. The scientists consider themselves normally detached from politics, but there are, as we will see, exceptions.

Since the beginning of this century there have been three relatively short periods in which the question of scientific responsibility, drew much attention both in scientific circles and outside the scientific community.

Two of the peaks of interest had to do with the role of the scientist in wartime. Hence shortly after the First and Second World War the subject became acute. The third wave came during the university reform period in the late sixties. Apart from these peaks, there is a steady increase of attention. This has to do with the fact that a growing percentage of scientists is employed by industrial companies, universities and research institutes.

As regards code of conduct there is a fundamental difference between scientists and other groups of professional academics, for instance physicians and lawyers, who have a reasonably well defined responsibility. Physicians have a direct responsibility to their patients and the medical professional organisations. Lawyers are responsible to either their clients or the courts. Since Galilei, scientists fostered a tradition to think and work free of prejudices, dogmas and taboos, defining their own responsibility (5). But nowadays most scientists have salaried functions. Hence, they have formally a direct responsibility to their employers. But in practice it is not as simple as that. This is because of the individualistic attitude of the scientists, based on the principle of scientific freedom.

During both world wars, particularly during the second war, scientists played an important role. In wartime the responsibility of the scientists was not openly discussed - they had to do what their country demanded. But after each war, discussions among scientists started concerning their responsibility and freedom. After the first World War the discussion came in a period when governments and big industrial companies started to incorporate science much more than before the war. Hence a rapidly increasing number of scientists got jobs outside the universities and other educational institutions. Not in all cases did they manage to maintain their scientific freedom.

Much more intensive were the discussions about scientific responsibility and scientific freedom after the Second World War, because that was to quite some extent a scientific war. The most striking example was the role of the nuclear physicists, culminating in the dropping of atomic bombs on Hiroshima and Nagasaki in 1945. After the war the politicians had to decide whether or not to continue the research on atomic weapons. It led to fervent discussions in scientific circles and to important statements by politicians. Remarkably enough the discussion was mainly restricted to atomic weapons.

41

According to Harry Hall (3), congressmen like most laymen, had little reason to be much concerned with either science or scientists until the Hiroshima bomb fell. "It threw science and scientists into the forefront of politicians' focus of attention. For the first time, senators and representatives found themselves in considerable interaction with scientists. Whether congressmen liked it or not, they had to take note of the crucial role of science and scientists in the atomic age."

Harry Hall mentions that the senators were confronted with the fact that the subject matter with which scientists dealt was, and always would be, a mystery to them. Politicians were permanently barred from gaining access to and knowledge of scientific matters. Consequently, in this one respect, they looked upon scientists as a sort of secret society from which they were excluded. One of the consequences was that the politicians were inclined to consider the progress of science as a major social force with a dynamic and progressively expanding character. And they realized that the role and impact of science would grow even greater with time.

Eventually, in some fields of science, a new type of scientists emerged. His or her scientific fervour goes hand in hand with a socially and politically oriented fervor. In retrospect, it appears that this was almost bound to occur.

## The scientist as activist

The first post-war example of scientists in the function of activists was the opposition against nuclear weapons and later also against the civilian use of nuclear energy. It has led many countries to a deviation from previously established unbiased scientific judgment as a basis for national energy policy. Examples are Sweden and The Netherlands.

The scientist-activist type is predominant in all subjects related to the environment, where global climate change has been a fashionable subject since the end of the sixties. At that time many meteorologists were convinced that we were approaching a new ice-age. Between 1970 and 1975 several alarming books on this threat appeared. In the media, interviews with well-known climatologists and meteorologists confirmed their conclusion that the people should be prepared for a period of severe winters and cool summers.

In the course of the seventies, however, a few American climatologists managed to convince their colleagues that the time had come to pay attention to another threat: the fear of a global warming in the next century, due to a manmade increase of carbon dioxide and other minor constituents in the atmosphere. They started from the hypothesis that an increase of the atmospheric concentration of the components responsible for the natural greenhouse effect would lead, via an increased greenhouse effect, to global warming. Their predictions concerning the magnitude of such an effect were based on the in these days available and fairly primitive computerized climate models. Nevertheless, they managed to obtain the support of two powerful UN-bodies, the WMO (World Meteorological Organization) and UNEP (United Nations Environment Programme).

Representatives of the WMO also persuaded the International Council of Scientific Unions (ICSU) that climate research should become one of its specific fields of attention, and consequently representatives of WMO, UNEP and ICSU met in Villach (Austria) in October 1985 to discuss this newly discovered threat to mankind. Seldom has a scientific meeting had such an influence on the policy making (4). The participants concluded that,

> "although quantitative uncertainty in models results persist, it is highly probable that increasing concentrations of the greenhouse gases will produce significant climatic change".

The conference statement mentioned that,

> "the understanding of the greenhouse question is sufficiently developed that scientists and policy-makers should begin an active collaboration to explore the effectiveness of alternative policies and adjustments".

It recommended that UNEP, WMO and ICSU take action to

> "initiate, if deemed necessary, consideration of a global convention".

The conclusions of the Villach Conference served also as the main source of information as regards climate change for the the World Commission on Environment and Development (WCED). This commission was called into existence by the General Assembly of the UN in 1983. Its main task was "to propose long-term environmental strategies for achieving sustainable development by the year 2000 and beyond". Its report "Our Common Future", published (8) in 1987, is usually named the Brundtland report, after the Norwegian prime minister and former environment minister Gro Harlem Brundtland, who chaired the WCED.

The Brundtland report contributed to spreading the idea that global warming was a major problem facing mankind. Its foreword mentioned:

> "Scientists bring to our attention urgent but complex problems bearing on our very survival: a warming globe, threats to the Earth's ozone layer, deserts consuming agricultural land".

43

The Brundtland report states that,

> "the greenhouse effect, one such threat to life-support systems, springs directly from increased resource use".

And it concludes, based on this brief reference to the Villach meeting that global warming

> "could cause sea-level rise over the next 45 years large enough to inundate many low-lying coastal cities and river deltas. It could also drastically upset national and international agricultural production and trade systems".

In June 1988 Canada sponsored an international conference on "The Changing Atmosphere: Implications for Global Security". The conference focused on climate change. One of its principle purposes was to bridge the gap between scientists and policy-makers. It was attended by 340 individuals from 46 countries, including two heads of state, more than 100 government officials, scientists, industry representatives and environmentalists (9).

The Toronto Conference issued a statement which began with the portentous words:

> "Humanity is conducting an unintended, uncontrolled, globally pervasive experiment whose ultimate consequence could be second only to a global nuclear war. It is imperative to act now".

As initial action, the Conference recommended:

- a 20 percent reduction in global $CO_2$ emissions from 1988 levels by the year 2005,

- development of "a comprehensive global convention as a framework for protocols on the protection of the atmosphere," and

- establishment of a "World Atmosphere Fund" to be financed in part by a levy on fossil fuel consumption in industrialised countries.

The Toronto Conference was not officially government-sanctioned, because the government participants attended in their personal capacities. Hence the Conference Statement was not binding on anyone. It was drafted by a committee composed mostly of environmentalists and discussed in less than a day and environmental activists dominated the scene. Many participants did not fully appreciate the political difficulties of addressing the climate change issue.

The attitude in governmental circles towards the conclusions of the Villach and Toronto meeting varied from enthusiastic approval in some countries to suspicion in others. Some officials considered the conclusions were the result of environmental activism rather than sound science. In an attempt to bring back governmental control to an increasingly important political issue, a number of governments had already requested WMO and UNEP to establish the Intergovernmental Panel on Climate Change with the mandate of providing "internationally co-ordinated assessments of the magnitude, timing and potential environmental and socio-economic impact of climate change and realistic response strategies". Thus it was that the year of the Toronto Conference (1988) also saw the birth of the IPCC.

## The influence of individual scientists

Another important event in 1988 was a hearing of the US Senate Energy Committee on the greenhouse effect. It was a good example of the unprecedented role played by some individual scientists in influencing policy-makers.

In the summer of 1988 the US suffered from a heatwave. The testimony of James Hansen, a NASA climate modeler, to the Senate Energy Committee that "the greenhouse effect has been detected and it is changing our climate now" made front page news, coming as it did at the height of the heatwave. Senator Max Baucus commented:

> "I sense that we are experiencing a major shift. It's like a shift of tectonic plates".

James Hansen should have known better because every meteorologist knows that regional heatwaves have nothing to do with climate change. They occur in a region when the distribution of high and low pressure centres shows little change for a number of weeks. During such a period the jet stream in the upper atmosphere hardly changes. Such a "blockage" seems to be a chaotic phenomenon.

In The Netherlands there have been thirty heatwaves so far in this century. According to the Royal Netherlands Meteorological Institute, there is no explanation for the puzzling occurrence of a heatwave. American meteorologists could have confirmed this. But thanks to Hansen's testimony, climate change burst on to the political stage in the US.

Even George Bush used it immediately in his election-speeches a few months later.

> "Those who think we are powerless to do anything about the greenhouse effect are forgetting about the White House effect. As President I intend to do something about it".

After the election, Chief of Staff John Sununu convinced the new President to adopt a different view.

This pressure on governments and the UN had two immediate results. It strongly influenced the discussions in the World Commission on Environment and Development and it initiated the formation in 1988 of the Intergovernmental Panel for Climate Change (IPCC), a joint venture of the WMO and the UNEP, set up as "an intergovernmental mechanism aimed at providing the basis for the development of a realistic and effective internationally accepted strategy for addressing climate change". The term mechanism, with its relationship to machinery, is quite appropriate to describe what has happened since 1988. Two IPCC reports (6,7), published in 1990 and 1992, became the cornerstone of a UN action which assumed climate change to be the great threat of the next century.

## The activities of the IPCC

The IPCC held its first meeting in November 1988, electing professor Bert Bolin (Stockholm) as chairman and establishing three working groups. Working Group I on science, chaired by the United Kingdom, Working Group II on impacts, chaired by the Soviet Union and Working Group III on response strategies, chaired by the United States. On December 6, 1988 the General Assembly of the UN adopted the establishment of the IPCC, urging governments, inter-governmental and non-governmental organisations, and scientific institutions to treat climate change as a priority issue. This decision triggered an escalation of speculative scientific thinking. In spite of many uncertainties in the underlying scientific theory and in spite of the deficiencies in the computer models used for climatic predictions, a small but vociferous group of scientists claimed that there was sufficient evidence to predict disastrous climatic developments for the next century.

The IPCC report "Climate Change" served as the major basic document for the Second World Climate Conference, convened in Geneva late in 1990. Not only WMO and UNEP, but also ICSU, UNESCO, FAO and the Intergovernmental Oceanographic Commission (IOC) acted as sponsors of this conference. Although the scientific discussions during that conference clearly demonstrated the many uncertainties, the organizers tried in vain to obtain the approval of the participants of a drafted conference statement in a final plenary session. A chaotic discussion resulted, but nevertheless a small steering group reported to the Environment Ministers (who met immediately after the Climate Conference) that the participants came to unanimous conclusions. And the ministers and other representatives from 137 countries ended their meeting with a Ministerial Declaration, expressing their serious concern about the newly discovered threat facing mankind.

What happened since then can be summarized as "more of the same". It led via numerous IPCC conferences and the UN Conference on Environment and Development (Rio de Janeiro 1992) to three sessions of the "Conference of the Parties to the United Nations Framework Convention on Climate Change" and to a gradual phasing out of the scientists. During the third (Kyoto, December 1997) they played only a minor role - the politicians had now taken the lead.

## Policy making by scientific consensus

A new and dangerous principle has been introduced in science based policy making, namely decisions based on scientific consensus by majority vote, achieved by a group of scientists. This gives rise to a number of crucial questions. If politics rests on "scientific consensus", what about democracy? Who is in charge? Who is responsible? How do we know that what is proclaimed to be a "consensus" really is a consensus view. And how do we know that this view comes anywhere near the truth?

In the case of the international politics of climate change, the preference of the IPCC for scientific consensus has recently been criticized by Hansson and Johanneson in their contribution to the book (2) "International Politics of Climate Change":

> "In many, perhaps most, cases, the search for scientific consensus is both reasonable and useful. In relation to the science of climate change, however, a too consensus-oriented approach may hide away one of the most important aspects that decision-makers need to take into account, namely scientific uncertainty. In order to provide decision-makers with as adequate information as possible, diverging opinions as well as majority opinions should be presented. Committee procedures for the IPCC and other similar scientific bodies remain to be developed that encourage rather than discourage the public presentation of minority opinions. This is particularly important in the case of the IPCC, since the vast majority of the worldís leading experts in the relevant fields are involved in the IPCC process."

47

Their plea for attention to a diversity of opinions should have been taken to heart. Unfortunately those scientists who do not whole-heartedly subscribe to the IPCC conclusions and recommendations are invariably characterized as ignorant outsiders or even worse, as lobbying in the interest of the coal and petroleum industries. They are referred to as a small minority of "contrarians". But science does not proceed by majority vote. Nor should it really matter in whose interest the scientist works. The only thing that counts is the strength of the arguments. Moreover, most of the dissenters as regards the IPCC are independent scientists far remote from the coal and oil business.

## Peer review

The IPCC has introduced another dangerous principle: only peer-reviewed publications are taken into consideration for the composition of their reports.

Peer review is an indispensable part of science. The procedure is that papers submitted for publication in a scientific journal are scrutinised by referees appointed by an impartial editor. But the IPCC procedure for peer review is different. The IPCC also takes into consideration papers not submitted by a scientist, but solicited by its own organisation. And the editors (solicitors) cannot be impartial because they are personally involved as "lead authors". They themselves appoint the peers.

Unfortunately the IPCC attitude towards peer review has been followed by the authors of the aforementioned Scandinavian book. This only partly explains why none of the dissenting Scandinavian publications are mentioned in the bibliography list of 500 references. Several of these relevant papers (10 - 16) have undergone more peer review than most of the 500 cited papers.

A final remark: in a relatively new field of science, such as climatology, peer review is in hands of a small inner circle. This can easily lead to refusal of papers not supporting the majority opinion.

## References

1. *Negotiating Climate Change: The Inside Story of the Rio Convention.* Edited by I.M. Mintzer and J.A. Leonard; Cambridge University Press 1994.
2. *International Politics of Climate Change, Key Issues and Critical Actors.* Edited by Gunnar Fermann. Oslo: Scandinavian University Press, 1997.
3. Hall, H. *The Sociology of Science.* Edited by Bernard Barber and Walter Hirsch; The Free Press of Glencoe, 1962.

4. Böttcher, C.J.F. *Science and Fiction of the Greenhouse Effect and Carbon Dioxide*, published by The Global Institute for the Study of Natural Resources, The Hague 1992.

5. Koestler, A. *The Sleepwalkers*, second ed. Pelican Books, London 1977: p. 358.

6. *Climate Change: The IPCC Scientific Assessment*. Edited by J.T. Houghton, G.J. Jenkins and J.J. Ephraums, Cambridge University Press 1990.

7. *Climate Change 1992: The Supplementary Report to the IPCC Scientific Assessment*. Edited by J.T. Houghton, B.A. Callander and S.K. Varney, Cambridge University Press 1992.

8. World Commission on Environment and Development (WCED); *Our Common Future*, Oxford University Press 1987.

9. Böttcher, C.J.F. *'Climate Change: Forcing a Treaty'*, Energy & Environment Vol. 7 (1996): p. 381.

10. Jaworowski, Z.; Hisdal, V. and Segalstad, T.V. *Atmospheric $CO_2$ and Global Warming: a critical review*, Norsk Polarinstitutt, report 119 (1990, second revised edition 1992).

11. Jaworowski, Z; Segalstad, T.V. and Ono, N. *Do Glaciers Tell a True Atmospheric CO2 Story?*, The Science of the Total Environment 114 (1992) 227.

12. Jaworowski, Z. *Ancient Atmosphere - Validity of Ice Records*, Environment Science and Pollution Research 1 (1994) 161.

13. Karlén, W.; Friis-Christensen, E. and Dahlström, B. *The Earth's Climate - natural variations and human influence*, Elforsk AB, Stockholm, 1993.

14. Gerholm, T.G. *Energy Demand and Climate Change*, Swedish Member Committee of the World Energy Council, 1995.

15. Friis-Christensen, E. and Lassen, K., Science 254 (1991) 698.

16. Svensmark, H. and Friis-Christensen, E., Journal of Atmospheric and Solar-Terrestrial Physics, 1997.

# Natural climate changes

# 4.

# Is the temperature increase of the last 100 years unique?

*Wibjörn Karlén*

The global temperature has increased during the last century. In order to determine if this change is unique, as often is claimed, and to what extent it may be a result of the release of greenhouse gases by mankind, it is necessary to know how the climate has varied over an extended period. Weather observations have been made and journals have been kept for many centuries, but it was first in 1659 in England that direct measurements of temperature were made. Temperature observations did not become common until the latter half of the 1800s. Therefore, knowledge concerning changes in the climate must largely be based on information obtained by indirect methods.

## Methods for reconstruction of past climate

A number of methods may be used to determine variations in past climate. These methods include, for instance, studies of changes in vegetation distribution, in the size of glaciers and of the water level in lakes without outlets (Lamb 1972, Bradley 1985). Some of the most detailed information has been obtained from studies of tree rings (Schweingruber 1996) and of variations in the content of stable isotopes in ice-cores obtained from polar ice sheets, corals and stalagmites (Dansgaard 1981, Lorius 1989, Jonsen et al. 1995, Alley et al. 1997, Holmgren et al. 1999). The accuracy of any reconstruction of the climate is highly dependent on good quality dating.

## Interpretation of paleoclimatic data

Paleoclimatic data is expressed in a number of different units, such as $\delta^{18}O/^{16}O$ (variations in the relation between the two oxygen isotopes in ice-cores, corals and stalagmites); mm (the width of tree rings); m (the altitude of the tree-limit and lake levels). To determine what climate history the parameters indicate, the record is compared with present day conditions. It is essential that the data series overlap with the period of temperature and precipitation observations and that several data sets can be compared. The qualitative information about changes in climate is usually reliable but quantitative estimates may be uncertain. Dating based on annual growth normally yields dates with only minor errors, whereas dates based on radioactive decomposition may be subject to errors of a few hundred years. The use of long time-series will require some form of smoothing. This, in combination with the use of statistical methods, may influence the results. Consideration must be given to these facts when paleoclimatic reconstructions are compared.

## Local or global changes in the past climate

The global temperature increase during this century has been determined using data from climatic stations throughout the world. Although the extent of this global warming is open to discussion, its existence cannot be called in question. This temperature increase is frequently considered unique, but paleoclimatologists point out that similar changes have occurred repeatedly. Advocates of a strong greenhouse effect argue that earlier changes in climate have been recorded only at a relatively small number of localities and therefore the reconstructed changes could be a result of the redistribution of heat. Only if it can be proved that changes in climate have occurred at a large number of localities simultaneously, will it be possible to prove that a change in the radiation balance of the planet has taken place. The occurrence of global changes in the climate is of central importance for the $CO_2$ discussion.

Departures from the long-term trend in global temperature during the period of instrumental observations have been documented (WMO 1996). Such deviations took place around 1880, 1910, 1940 and 1975.

The Holocene period (the last 10 000 years) has often been considered a period of basically stable climate. As long as no marked variations were recognised, there was little reason to discuss correlations between areas. However, it is now known that the climate has fluctuated even during the Holocene and, thus the timing of these fluctuations has become a question of central importance for the understanding of the processes that determine climate (Karlén &

53

Kuylenstierna 1996, Alley et al. 1997, Bond et al. 1997, Bianchi & McCave 1999, Holmgren et al. 1999). The question of whether there have been concurrent global climatic changes before this century has been discussed, but too little is known to permit a conclusive answer (Alley et al. 1997, Kreutz et al. 1997). However, the relationship between El Niño and La Niña currents in the Pacific Ocean with global climate supports the view that a global pattern in climate does occur (WMO 1999).

## Climate data for the past forty years

For the past forty years the global change in temperature has been determined using several independent data series (Ahlbeck, this volume). P.D. Jones (WMO 1995) has compiled the data on which the most frequently discussed view of global temperature is based and it is on this data set that the the IPCC reports are focused (Houghton et al. 1990, 1992, 1996). This global temperature series is based on observations from a large number of meteorological stations throughout the world. Other data series have been compiled from balloon and satellite observations (Christy & McNider 1994, WMO 1997, Gaffen 1998) (Fig. 4.1). Short-term variations in the global temperature occur at the same time in all three data sets. However, while P.D. Jones's time series shows a marked and still ongoing increase in temperature beginning around 1975, the time series based on balloon and satellite data show no or only a small trend after 1980 (WMO 1997).

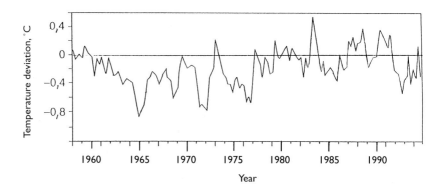

**Figure 4.1.** The global temperature for the last decades has been estimated from traditional ground measurements as well as from balloon and satellite data. According to the balloon data the global temperature increased between 1975 and 1980 but thereafter no trend is recorded. The temperature estimate based on ground observations indicates that the global temperature has continued to increase (Kerr 1995).

There is no accepted explanation for the difference between the temperature series. It has been suggested that it depends on an error in the mathematical treatment of the satellite data (Gaffen 1998), but this does not explain why the balloon data observations do not show the temperature increase during recent years that the surface data indicate. To some extent the difference may be a result of a marked temperature increase during the fall, winter and spring in Siberia, which would add somewhat to the global mean temperature. Because satellite and balloon observations record the temperature in the lower atmosphere, a change in the frequency of extremely cold ground temperatures will not be recorded in these data sets (WMO 1995).

Several explanations that are not related to the increased atmospheric content of greenhouse gases have been discussed for the marked increase in temperature in Siberia during recent years. For example, it has been suggested that several stations could be affected by urban warming. A second explanation is that there has been a decrease in the frequency of temperature inversions and extrreme low temperatures at ground level, which is a result of radiation cooling.

## Temperatures in Europe since the 1600s

For the period prior to the mid 1800s, there are far too few places where temperature observations were made to allow an estimate of global temperature with the same degree of accuracy as that of the last century. However, there are several temperature observation series that were initiated during the late 1600s and early 1700s (Fig. 4.2). Because the temperature trends of these stations are similar to the global trend of the last century, it is likely they do give a general idea about northern hemisphere temperature. This is illustrated in figure 4.3 where the temperature for the northern hemisphere is compared with temperature data from Uppsala (Moberg and Bergström 1997).

The data series from both England and Uppsala indicate that the climate of the 1730s was about as warm as that of the 1930s and of recent years (Fig. 4.2). The De Bilt (Holland) series indicates that the temperature in the 1730s was actually warmer than that of the 1930s. However, the temperature around 1940 in the De Bilt series is not quite as high as the temperatures recorded during the past ten years. Over this three hundred-year period as a whole, the trend towards a warmer climate is small, even if the temperature increase since late 1800s is distinct.

Bradley and Jones (1993) have presented temperature observations beginning during the 1700s or early 1800s for ten places in the northern hemisphere. The temperature changes shown are similar to the one in the long temperature series mentioned above. Typically, the climate seems to have been

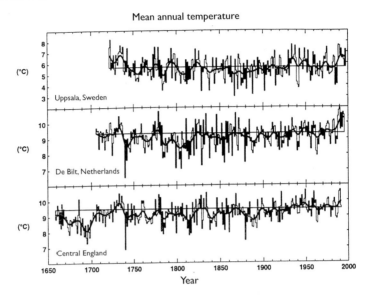

**Figure 4.2.** Direct observations of annual average temperatures for the three longest observation series in the world. These three series from western Europe indicate a similar trend and because it is similar to the global temperature during the overlapping period, it is likely that these series yield a general view of northern hemisphere temperature. There is a clear trend in the temperature during this century. However, if the whole period is considered, the trend is minor (Moberg and Bergström 1997).

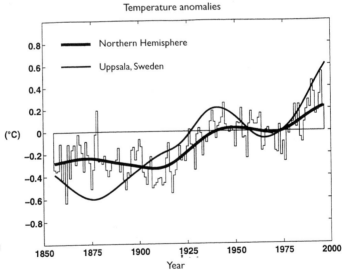

**Figure 4.3.** The deviation from normal annual temperature for the northern hemisphere and for Uppsala, Sweden. The temperature record from Uppsala, one of the places for which observations are available since early 1700s, shows changes in the temperature trend at approximately the same time as the temperature trend changed for the northern hemisphere. The temperature increased from the turn of the century and up to about 1940. Thereafter it decreased until about 1975, When it again began to increas (WMO 1995, Moberg and Bergström 1997).

56

relatively warm during the late 1700s, warm around 1830, cool during the late 1800s and warm between 1930 and 1950. Although some stations show a distinct trend towards a warmer climate, there is little or no trend apparent for other stations. The urban effect must be considered when information from towns and airports is used. An extensive study carried out in USA has revealed that the effect of urbanisation is greater than is often assumed (Jones et al. 1989, Karl and Jones 1989).

Relatively detailed information is also available for the 1600s. These data series, which are based on dendrochronologies and ice-cores (Houghton 1992), show that the climate was cold during most of the1600s. The same trend is also evident from an analysis of 29 data series covering the Arctic area of the northern hemisphere (Overpeck et al. 1997). Several data series from the Arctic area indicate a temperature decrease after the 1950s.

## Changes in the climate during the Holocene

Although it has generally been assumed that Holocene climate has been stable (Broecker 1994, Johnsen et al. 1995, Bond et al. 1997), however, recent methods permitting a better resolution than it was possible to obtain only a few years ago show that even the Holocene climate has been subject to variations. The range of these variations has been around 2°C (e.g. Lorius et al. 1979, Bonnefille et al. 1990, Lin et al. 1995, Karlén & Kuylenstierna 1996, Alley et al. 1997, Wick & Tinner 1997). This is small in comparison with variations during the last Ice Age, when the range of temperature variations over a few millennia was about 20°C on Greenland (Johnsen et al. 1995), although the average global temperature decrease probably only was 5-6°C.

It was first when O'Brien et al. (1995) found sizeable variations in the content of soluble salts in ice-cores from Greenland that the concept of major, possibly global, changes in the atmospheric circulation gained acceptance (Bond et al. 1997). These variations are of the same duration as the well-known, large-amplitude variations during the last glaciation, but the amplitude of the Holocene variations is much smaller (Johnsen et al. 1995, Karlén & Kuylenstierna 1996, Bond et al. 1997). Extensive studies of major climatic changes that occurred at the end of the last Ice Age lend support to the view that large climatic fluctuations have taken place at the same time over large areas (Alley et al. 1997, Broecker 1997, Severinghaus et al. 1998, Fischer et al. 1999).

# Forcing

Processes that can cause regional changes in climate have been discussed. One such process is the variation that occurs in the solar irradiation. DeVries (1958) first suggested a relation between ¹⁴C production, caused by the modulation of solar radiation, and climate. Denton & Karlén (1973) suggested that reduced solar irradiation could be the forcing mechanism which caused glacier expansions in several widely separated areas approximately every 2500 years during the Holocene. This view received support from an extensive study of worldwide glacier fluctuations by Röthlisberger (1986), although Grove (1988) argued that there only was evidence for global changes of this type during the Little Ice Age period (the last five hundred years). Later studies have demonstrated a close correlation between changes in the Holocene climate and solar variability (Wigley & Kelly 1990, Karlén & Kuylenstierna 1996) (Fig. 4.4).

The view that variations in solar irradiation could affect climate gained support when Friis-Christensen & Larssen (1991) demonstrated a close correlation between the length of the sunspot cycle and climate. Lean et al. (1995) and Soon et al. (1996) have modelled the effect of an estimated amount of solar activity on climate since AD 1610 and found that solar variability is likely to explain at least some parts of the observed variations in climate during the last several hundred years. However, the variations in solar irradiation are small and the effect must be reinforced by feedback mechanisms. There appears to be a correspondence between the change in the amount of cloud cover appear to vary with the number of sunspots, and this is thus one possible feedback mechanism (Svensmark & Friis-Christensen 1997).

Index showing variations in solar irradiation

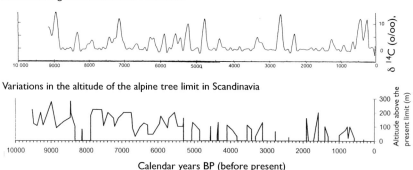

Variations in the altitude of the alpine tree limit in Scandinavia

Calendar years BP (before present)

**Figure 4.4.** The alpine tree limit in Scandinavia has fluctuated with changes in the summer temperature. Variations in the alpine tree limit are shown here together with variations in solar radiation, which have been calculated from the difference between dendrochronological age and the ¹⁴C age of tree rings (Wigley and Kelly 1990). Most distinct periods of low alpine tree limit coincide with periods of low solar irradiation.

The solar impact on climate is further supported by new evidence showing periodic changes in the dust content of the Greenland ice sheet. Periods with lengths of 11 years, 91 years and ~200 years, which are periods that correspond to variations in solar activity, are found in a study of a section of the GISP2 ice-core covering almost 90 000 years (Ram and Stolz 1999).

Global changes in the climate can also be a result of major volcanic eruptions (Sear et al. 1987, Bryson & Bryson 1997, Overpeck et al. 1997). The effect of recent volcanic eruptions has been well established by surface temperature observations as well as by balloon- and satellite observations (Sear et al. 1987). Forcing appears to be global for volcanic eruptions taking place in the southern hemisphere as well as at low latitudes, while eruptions taking place in the northern hemisphere mainly affect this hemisphere (Sear et al. 1987). The effect of a single eruption is restricted to a few years (Hammer et al. 1980, Zielinski et al. 1996); when several eruptions occur during one period the forcing can affect the climate in such a way that the global temperature will be lowered during an extended period. The near absence of volcanic eruptions during the first half of this century is a factor likely to have contributed to the temperature increase during this period.

## Contemporaneous change in climate, examples

Dendrochronological studies from the northern hemisphere show that major climatic changes occurred at approximately the same time in several places. Studies from northern Sweden (Fig. 4.5) and from the Rocky Mountains indicate particularly warm events around AD 1100, 1160, 1320, 1430, 1760 and 1940. The warming that has taken place during this century is pronounced but not more distinct than the events that took place during the 1300s, the early 1400s and in the late 1700s (Briffa et al. 1992, Luckman et al. 1997).

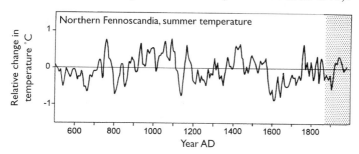

**Figure 4.5.** This tree ring series from Lapland, northern Sweden, shows that distinct variations occurred in the summer temperature on a number of occasions before the 20th century (Briffa et al. 1992). Statistics of extreme climate in Scandinavia during the past 1500 years for a) individual years, b) 20-year averages, c) 50-year averages and d) 50-year trends indicate that the temperature changes that have occurred during this century are not unique.

As mentioned before, the number of well-dated temperature series covering an extended period is limited. However, there are several data sets that show that at least major changes in the climate took place at approximately the same time in localities located far apart. Examples of this type are the Greenland ice-core (Alley et al. 1997), and the Scandinavian pine tree limit (Karlén and Kuylenstierna 1996). These two data sets also show variations during periods that correspond to those indicated by dates obtained for variations in the production of North Atlantic Deep Water (Bianchi & McCave 1999). In addition, there is a relatively close similarity between data from the same Greenland ice-core and data on cave temperature in South Africa (Fig. 4.6).

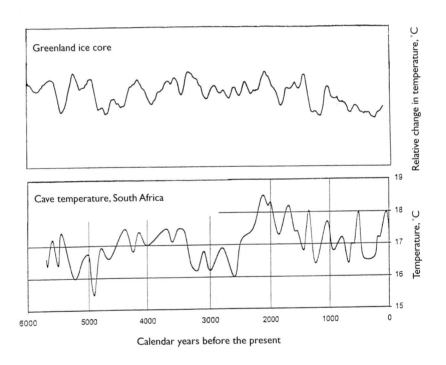

**Figure 4.6.** Estimated variations in the temperature for Greenland (Alley et al. 1997) and a reconstructed record of temperature variations for a cave in South Africa (Talma & Vogel 1992). The records are similar during the last 2000 years. Before that there seems to be an error in the time scale of up to 200-300 years around 5000 years ago. The low temperature in the South African cave 4900 years before the present and the long cool period 3300-2400 years ago are not distinct in the Greenland ice core but are distinct in the Swedish chronology. The warm period in South Africa 500 years ago coincide with a marked warm period in Scandinavia.

## Conclusions

- The global temperature for the last 100 years is known from direct observations. The temperature was low until 1910; thereafter it increased up to 1940. After a period of slightly lower temperature around 1975, the global temperature has increased.

- The longest temperature-observation series from England, Holland and Sweden, show a similar pattern of changes in climate. For the timespan during which these series overlap with the estimations of global temperature (since 1850) there are similarities. It is therefore likely that these three temperature-observation series provide a reasonable description of the climate of the northern hemisphere. There is a slight trend indicating an increase in temperature during the last three centuries. However, several changes in temperature of almost the same magnitude as between 1920 and 1940 have occurred.

- There are also records about changes in the climate during the last 1000 years from dendrochronologies, ice-cores and speleothems. Changes spanning several decades appear to occur at the same time in several well-dated chronologies, although short-term variations can be out-of-phase. The temperature changes that have occurred during this century do not appear to be unique.

- There are a few well dated climatic series that provide data for the entire Holocene. These include chronologies based on ice-core data from Greenland and Antarctica as well as chronologies based on tree-limit variations, glacier fluctuations and variations in lacustrine sediments from arid environments. In addition, there are a large number of chronologies that cover parts of the Holocene. These studies indicate that major changes in climate occur at approximately the same time.

- The temperature change in degrees centigrade can be estimated with a varying accuracy. In general, reconstructed changes in climate are given as an average of a 10 to 100 year periods. Dendrochronologies from western USA and Sweden as well as data from studies of tree-limit variations in Sweden indicate variations of about 2°C, while data from stalagmites from southern Africa indicate slightly larger variations.

61

# References

Alley, R.B., Mayewski, P.A., Sowers, T., Stuiver, M., Taylor, K.C. and Clark, P.U. (1997), Holocene climatic instability: A prominent, widespread event 8200 yr. ago. *Geology* 25(6): 483-486.

Bianchi, G.G. and McCave, I.N. (1999), Holocene periodicity in North Atlantic climate and deep-ocean flow south of Iceland. *Nature* 397: 515-517.

Bond, G., Showers, W., Cheseby, M., Lotti, R., Almasi, P., deMenocal, P., Priore, P., Cullen, H., Hajdas, I. and Bonani, G. (1997), A pervasive millenial-scale cycle in North Atlantic Holocene and Glacial climates. *Science* 278: 1257-1266.

Bonnefille, R., Roeland, J.C. and Guiot, J. (1990), Temperature and rainfall estimates for the past 40,000 years i equatorial Africa. *Nature* 346: 347-349.

Bradley, R.S. (1985), *Quaternary Paleoclimatology, Methods of Paleoclimatic Reconstruction*. Allen & Unwin, Boston, 472 p.

Bradley, R.S. and Jones, P.D. (1993), 'Little Ice Age' summer temperature variations: their nature and relevance to recent global warming trends. *The Holocene* 3(4): 367-376.

Briffa, K.R., Jones, P.D., Bartholin, T.S., Eckstein, D., Schweingruber, F.H., Karlén, W., Zetterberg, P. and Eronen, M. (1992), Fennoscandian summers from A.D. 500: Temperature changes on short and long timescales. *Climatic Dynamics* 7: 111-119.

Broecker, W.C. (1994), Massive iceberg discharges as triggers for global climate change. *Nature* 372: 421-424.

Broecker, W.S. (1997) Thermohaline circulation, the Achilles heel of our climate system: Man-made $CO_2$ upset the current balance? *Science* 278: 1582-1588.

Bryson, R.A. and Bryson R.U. (1997), Macrophysical climatic modeling of Africa's Late Quaternary climat: Site-specific, high-resolution applications for archaeology. *African Archaeological Review* 14(3): 143-160.

Christy, J.R. and McNider, R.T. (1994), Satellite greenhouse signal. Nature 367: 325.

Dansgaard, W. (1981), Paleo-climatic studies on ice cores. In, Berger, A. (Ed.): *Climatic variations and variability: Facts and theories*. D. Reidel Publishing Company, Dordrecht: Holland (795 p.): 193-206.

Denton, G.H. and Karlén, W. (1973), Holocene climatic variations - their pattern and possible cause. *Quaternary Research* 3(2): 155-205.

DeVries, H.L. (1958), *Variation in concentration of radiocarbon with time and location on Earth*. Koninkl Nederlandse Akadd Wetenschappen, Proc. Ser. B., 61: 94-102.

Fischer, H., Wahlen, M., Smith, J. Mastroianni, D. and Deck, B. (1999),- Ice core records of atmospheric $CO_2$ around the last three glacial terminations. *Science* 283: 1712-1714.

Friis-Christensen, E. and Lassen, K. (1991), Length of the solar cycle: An indicator of solar activity closely associated with climate. *Science* 245: 698-700.

Gaffen, D.J. (1998), Falling satellies, rising temperatures? *Nature* 394: 615-616.

Grove, J.M. (1988), *The Little Ice Age*. London Methuen, 498 p.

Hammer, C.U., Clausen, H.B., and Dansgaard, W. (1980), Greenland ice sheet evidence of post-glacial volcanism and its climatic impact. *Nature* 288: 230-235.

Holmgren, K., Karlén, W., Lauritzen, S.E., Lee-Thorp, J.A., Patridge. T.C., Piketh, S., Repinski, P., Stevenson, C. Svanered, O. and Tyson, P.D. (1999), A 3000-year high-resolution record of palaeoclimate for north-eastern South Africa. *The Holocene* 9(3): 295-309.

Houghton, J.T., Jenkins, G.J. and Ephraums, J.J. (1990), *Climatic change. The IPCC scientific assessment*. World Meteorological Organization/United Nations Environment Programme, Cambridge University Press, 365 p.

Houghton, J.T., Callander, B.A. and Varney, S.K. (1992), *Climatic change. The supplementary report to the IPCC scientific assessment*. World Meteorological Organization/United Nations Environment Programme, Cambridge University Press, 200 p.

Houghton, J.T., Meira Filho, L.G., Callander, B.A., Harris, N., Kattenberg, A. and Maskell, K. (1996), *Climatic change 1995. TheScience of Climate Change*. Cambridge University Press, 572 p.

Johnsen, S.J., Dahl-Jensen, D., Dansgaard, W. and Gundestrup, N. (1995), Greenland paleotemperatures derived from GRIP bore hole temperature and ice core isotope profiles. *Tellus* 47B: 624-629.

Jones, P.D., Kelly, P.M., Goodess, C.M. and Karl, T.R. (1989), The effect of urban warming on the Northern Hemisphere temperature average. *Journal of Climatology* 2: 285-290.

Karl, T.R. and Jones, P.D. (1989), Urban bias in area- averaged surface air temperature trends. *Bulletin American Meteorologist Society* 70(3): 265-270.

Karlén, W. and Kuylenstierna, J. (1996), On solar forcing of Holocene climate: evidence from Scandinavia. *The Holocene* 6(3): 359-365.

Kerr, R.A. (1995), Isthe World warming or not? Science 267: 612.

Kreutz, K.J., Mayewski, P.A., Meeker, L.D., Twickler, M.S., Whitlow, S.I. and Pittalwala, I.I. (1997), Bipolar changes in atmospheric circulation during the Little Ice Age. *Science* 277: 1294-1296.

Lamb, H.H. (1972), *Climate: Present, Past and Future - Volume 1: Fundamentals and Climate Now*. London, Methuen, 835 p.

Lean, J., Beer, J. and Bradley, R. (1995), Reconstruction of solar irradiance since 1610: Implications for climate change. *Geophysical Research Letters* 22: 3195-3198.

Lin, P.N., Thompson, L.G., Davis, M.E. and Mosley-Thmpson, E. (1995), 1000 years of climatic changes in China: Ice-core $\delta^{18}$ O evidence. *Annals of Glaciology* 21: 189-195.

Lorius, C. (1989), *Polar ice cores:A record of climatic and environmental changes.*In, Bradley(Ed.), Global changes of the past, p. 261-293, UCAR/Office for Interdisciplinary Earth Studies, Boulder, Colorado, 514 p.

Lorius, C., Merlivat, L., Duval, P., Jouzel, J. and Pourchet, M. (1979), Evidence of climatic change in Antarctica ovr the last 30 000 years from the Dome C icecore. Sea Level, Ice and Climatic Change, Proceedings of the Canberra Symposium, December 1979. *IHAS publication no. 131*, p. 217-225.

Luckman, B.H., Briffa, K.R., Jones, P.D., and Schweingruber, F.H. (1997), Tree-ring based reconstruction of summer temperatures at the Columbia Icefield, Alberta, canada, AD 1073-1983. *The Holocene* 7(4): 375-389.

Moberg, A. and Bergström, H. (1997), Homogenization of Swedish temperature data. Part iii: The long temperature records from Uppsala and Stockholm. *International Journal of Climatology* 17: 667-699.

O'Brien, S.R., Mayewski, P.A., Meeker, L.D., Meese, D.A., Twickler, M.s. and Whitlow, S.I. (1995), Complexity of Holocene climate as reconstructed from a Greenland ice core. *Science* 270: 1962-1964.

Overpeck, J., Hughen, K., Hardy, D., Bradley, R., Case, R., Douglas, M., Finney, B., Gajewski, K., Jacoby, G., Jennings, A., Lamoureux, S., Lasca, A., MacDonald, G., Moore, J., Retelle, M., Smith, S., Wolfe, A. and Zielinski, G. (1997), Arctic environmental change of the last four centuries. *Science* 278: 1251-1256.

Ram, M and Stolz, M.R. (1999), Possible solar influences on the dust profile of the GISP2 ice core from Central Greenland. *Geophysical Research Letters* 26(8): 1043-1046.

Röthlishberger, F. (1986), *10,000 Jahre Gletschergeschichte der Erde*. Verlag Sauerländer, 416 p. Aarau.

Schweingruber, F.H. (1996), *Tree rings and environment dendrochronology*. Paul Haupt Publishers Berne, 609 p.

Severinghaus, J.P., Sowers, T., Brook, E.J., Alley, R.B. and Bender, M.L. (1998), Timing of abrupt climate change at the end of the Younger Dryas interval from thermally fractionated gases in polar ice. *Nature* 391: 141-146.

Soon, W.H., Posmenties, E.S. and Baliunas, S.L. (1996), Inference of solar irradiance variability from terrestrial temperature changes, 1880-1993: An astronomical application of the sun climate connection. *Preprint Series No. 4344*, Harvard-Smithsonian center for Atrophysics, 26 p.

Svensmark, H. and Friis-Christensen, E. (1997), Variation of cosmic ray flux and global cloud coverage-a missing link in solar-climate relationships. *Journal of Atmospheric and Solar Terrestrial Physics* 59(11), 1225-1232.

Talma, A.S. and Vogel, J.C. (1992), Late Quaternary paleotemperatures derived from a speleothem from Cango Caves, Cape Province, South Africa. *Quaternary Research* 37: 303-213.

Wick, L. and Tinner, W. (1997), Vegetation changes and timberline fluctuations in the central Alps as indicators of Holocene climatic oscillations. *Arctic and Alpine Research* 29(4): 445-458.

Wigley, T.M.L. and Kelly, P.M. (1990), Holocene climatic change, $^{14}$C wiggles and variations in solar irradiance. *Philosopie Transactions Royalty Society London A* 330: 547-560.

WMO (1995), *The global climate system review. Climate System Monitoring June 1991-November 1993.* WMO-No. 819, Geneva, 150 p.

WMO (1996), *WMO Statement on the status of the global climate in 1995.* World Meteorological Organization WMO-No. 838, Geneva, 11 p.

WMO (1997), *WMO Statement on the status of the global climate in 1996.* World Meteorological Organization WMO-No. 858, 11 p.

WMO (1999), *WMO Statement on the status of the global climate in 1998.* WorldMeteorological Organization WMO-No. 896, 12 p.

Zielinski, G.A., et al. (1996), A 110,000-yr record of explosive volcanism from the GISP2 (Greenland) Ice core. *Quaternary Research* 45(2): 109-118.

# 5.

# Has the greenhouse effect changed the global climate?

*Jarl Ahlbeck*

An anthropogenically induced climatic change should either lead to a gradual increase or decrease in the frequency and intensity of extreme climatic events and /or a gradual, continual warming or cooling of the earth's atmosphere. However it has not been possible to establish that climatic variations have increased between different years. Nor is there concrete evidence to show that extreme climatic events have occurred more frequently during the twentieth century (IPCC 1996, p 173).

It is often stated that tropical cyclones have become more frequent and more intense and that the situation is likely to become worse in the future. Here we may quote the IPCC (1996, p. 134):

> "In conclusion, it is not possible to say whether the frequency, area of occurrence, time of occurrence, mean intensity or maximum intensity of tropical cyclones will change."

New studies on the other hand indicate that the frequency of storms and typhoons has slightly declined compared to previous decades (Landsea et al. 1996). Nevertheless, the damage of extreme climatic events may have increased due to non-climatic changes such as deforestation and destruction of natural water barriers.

## "The identification of fingerprints"

In the mass media, it has been stated that the balance of evidence indicates that the earth's climate has been subject to human influence (IPCC 1996) and that this is clearly demonstrated by the rise in global temperature caused by increased emissions of carbon dioxide in the atmosphere. However it is important to recognise that researchers have not been able to identify such an influence in the temperature data using normal scientific methods.

In order to understand how research into climatic change is conducted, I would wish to introduce a concept entitled the "identification of fingerprints". This type of research does not involve the use of traditional methods to find an anthropogenic signal in the climatic data but rather the application of computerised climatic models in order to estimate for example temperature and rainfall data for different regions and altitudes. Estimates are then made for the spatial correlation coefficients between the local model data and the observed data. An analysis is then made of changes in these correlation coefficients over time (Hegerl et al. 1997, Tett et al. 1996 and Santer et al. 1996). It is assumed that any year on year improvement in these correlations indicates that an anthropogenic signal has been identified.

An alleged improvement in the trend of these correlation coefficients was one of the reasons for the frequently quoted statement in the *Summary for Policymakers* regarding human interference with the climate. However the improvement in this correlation coefficient is not statistically significant for a longer period than that analysed by Santer et al. This indicates that the climatic models are unable to describe natural variations which in practice calls into question the identification of anthropogenic signals using this type of model.

The IPCC (p423) states that the most satisfactory results are obtained when anthropogenic factors such as aerosols and carbon dioxide are included in the model of climatic change. This is a well-known feature of advanced process models: the greater the number of parameters that are used in a model, the more satisfactory the results obtained, in spite of the physically nonsensical nature of many of the relationships. The improvement in the results may be due to an increased general flexibility of the model and not to a reliable quantitative description of the physical processes.

## Temperature time series in a specific region

*The separation of an anthropogenic trend from a natural trend*

In this section , I would wish to discuss temperature measurements and their interpretation using classical methods of evaluation. I have chosen this approach since in my view only one identifiable anthropogenic climate change, using classical methods of evaluation, is sufficiently large to have any practical significance for mankind.

67

With the exception of the summary, there are few grounds for disagreement with the direct evaluations of the temperature data in the IPCC report (1996). However in view of the important role played by the summary in the debate on these issues, it is necessary to cast a more critical eye on its interpretation of temperature measurements. Moreover new data and new interesting hypotheses have emerged since the report was written.

If one has access to an estimated natural trend, it is possible to assess whether or not the change in the observed measurement data deviates significantly from the natural trend. In statistical terms, it can be stated that if the natural trend lies outside the range of variation for the observed measurement data, it is likely that an anthropogenic component has been located.

### Separating an anthropogenic variation from a natural variation

The type of test outlined above cannot be carried out when the simulation programmes that we use are unable to describe natural trends. On the other hand if data is available for the range of variation of the natural trend, it will be possible to test whether or not the variation observed during the studied time period deviates significantly from natural variations.

The standard deviation can be used to measure of the range of variation of the natural trend . Here a very early period can be used for the measurement of average annual temperature in order to eliminate the influence of anthropogenic factors. It is important however to choose as large a time interval as possible in order to ensure that the measurements are not autocorrelated i.e. the temperatures for different years should not be interdependent.

### Identification of an anthropogenic trend if the natural variation is known

If the increase in the standard deviation of average annual temperatures is due to an unnatural (anthropogenic) warming or cooling trend, the possibility of using statistical methods to separate an anthropogenic trend from a natural trend requires that the increase (fall) in temperature is twice the natural standard deviation.

### Average temperatures in Finland 1902-1996

Let us now look at the case of average temperatures in a specific region. Here the annual average temperatures in Finland between 1900 and 1996 are chosen as an example. (Figure 5.1).

There is no distinct trend at all. On the basis of the data for this century, it seems therefore probable that the natural standard deviation of average temperatures in Finland is about 1°C. If we had access to data stretching back over previous centuries , the standard deviation would most likely be substantially larger.

The climatic models predict a regional warming for Finland amounting to several degrees (IPCC 1996) as a result of the doubling of the amount of carbon dioxide in the atmosphere estimated on the basis of a preindustrial level of 280 ppmv. A considerable proportion of this warming should already have taken place , particularly after the Second World War. However this is not borne out by the temperature data.

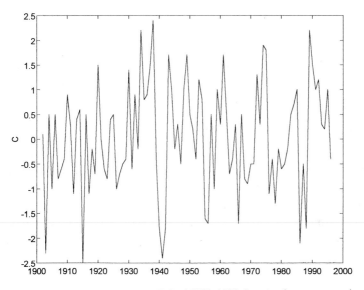

**Figure 5.1** Annual average temperatures in Finland 1900 -1996, deviation from average values.

# Horizontal and vertical average formation

*Global annual average temperatures, surface measurements*

The example shows that the natural variations in the studied region (Finland) are so large that it is not possible to identify an anthropogenic signal. A large degree of natural variance is a characteristic of local temperature data. These variations become greater, the further north that one goes.

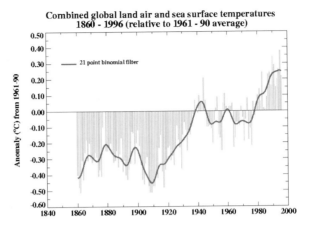

**Figure 5.2** Global average temperatures (surface measurements) 1860-1996, deviation from average values (The figures are published with the approval of Matt O'Donnel at the Hadley Centre, Meteorological Office, England. Dr. Phil Jones, Climate Research Unit, University of East Anglia, England has assisted with the analysis of land temperatures.)

In order to reduce the standard deviation, it is necessary to construct averages for much larger areas, even at a global level i.e. global average temperatures. Karlén in this book and Lindzen (1994) have discussed the substantial uncertainty inherent in these measures e.g. the observations from tropical oceans. This uncertainty reduces the reliability of such global climatic measures. Global average temperatures based on surface measurements from this century are presented in Figure 5.2.

*A comparison between global average temperatures and satellite measurements.*

Satellites measure the radiation emission of oxygen molecules in the night hemisphere. In this way a series of average air temperatures are produced for different altitudes. The measurement signals are tested with the help of weather balloons and the degree of agreement is almost perfect.

The precision and the resolution of satellite measurements is evident from the fact that we are even able to measure the heat given off by the moon. A full moon raises global temperatures by 0.03°C. The satellite measurements also show that the temperature at the weekends in Europe is 0.02°C lower which is presumably attributable to lower industrial activity and less traffic.

Satellite measurements have been carried out since 1979. An analysis of this data in comparison with surface data shows (Christy and McNider 1994) an acceptable measurement of agreement for land areas but somewhat poorer results for the oceans. This leads to a divergent trend for global average temperatures (Figure 5.3). This phenomena is discussed by Christy who corrects the satellite data to take account of the Mount Pinatubo volcanic explosion in 1992 and changes in ocean currents. Following these adjustments, Christy obtains a rising global temperature trend of 0.09°C in ten years.

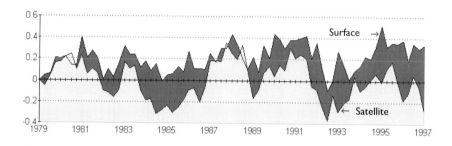

**Figure 5.3** Temperature anomaly (°C), surface and satellite (Daly 1998).

However Christy's correction was applied to data ending in 1993. However if we extend the data up until 1997 when the satellites continue to show no significant trend over the oceans despite the cessation of the Mount Pinatubo effect, the discrepancy continues to exist: satellite data fails to corroborate the weak trend towards global warming shown in the surface measurements.

A discussion on temperature data for the period 1965-1996 is to be found in the New Scientist for July 19th. 1997. During the first half of this period , 1965-1987, the extent of warming was lower in the northern hemisphere than in the southern. Climate research models explain this phenomena with reference to the cooling effect of aerosols in the northern hemisphere. In the subsequent period, 1987-1996, the northern hemisphere has warmed up more rapidly as a result of the triumph of the warming effect of carbon dioxide over the cooling effect of aerosols. The critics on the other hand have pointed out that average temperatures have not risen at all in the southern hemisphere but have actually fallen since 1987, which does not agree with the predictions of the model. Another fact that would also appear to contradict this explanation is that *night temperatures* in the northern hemisphere have risen, particularly in the winter.

The temperature increase at the beginning of the twentieth century, 0.5°C up until 1938, occurred at the same time as levels of carbon dioxide in the atmosphere were only 10 ppmv. This increase in global temperatures has presumably been natural and provides us with some idea of the extent of natural variations. However the global temperature curve is clearly subject to auto-

71

correlation which means that one ought to choose random samples at an interval of a decade. A high level of autocorrelation means that individual years with high or low temperatures tend to cluster. The temperature of individual years has no relevance when discussing climate change. The warm year, 1998, was due to the strong El Niño during 1997 and 1998. The even stronger El Niño in 1982 and 1983 did not increase the temperature as much as in the 1997/98 El Niño, because the dust emitted from the huge eruption of the El Chicon volcano in Mexico in 1982 caused a significant cooling of the atmosphere.

This suggests that one ought to have access to global average temperature data for periods of several hundred years in order to be able to estimate the natural standard deviation with any degree of accuracy.

Hence there is considerable uncertainty surrounding a statistical estimate of the extent of anthropogenic global warming that is required to ensure a statistically significant result. A global temperature increase of 0.2°C since 1938 or 0.3°C since 1965 cannot be said under any circumstances to deviate from the natural variations.

*Rising temperatures in the northern industrial belt between 1979-1995*

The north-south distribution of temperature trends (°C/10) based on data presented by Christy (1995) is depicted in Figure 5.4. The figure shows, as has been already described, the discrepancy between satellite measurements and surface measurements taken above southern oceans and around the equator. Both satellite and ground measurements indicate that a certain degree of warming has occurred in a belt between 40° and 65° north of the equator i.e. in the area where most of the northern hemisphere's industrial activity and international flights are concentrated. This rise in temperatures would appear to be particularly associated with winter nights. In Siberia, the surface measurements show remarkably large rises in temperature compared with the satellite measurements. The phenomena and causes of the rising temperatures in the northern industrial belt have been widely discussed. The explanations are hypothetical.

# Polar areas

## Melting glaciers

According to the climate models, the most marked incidence of global warming due to the greenhouse effect should occur at the poles. Hence temperature studies in the polar regions are of particular interest. Indirect indications of global warming such as melting glaciers are not as important as temperature measurements although they receive considerable attention in the mass media. The movement of glaciers is subject to considerable periodicity: if a glacier has moved particularly rapidly for a certain period, its lower section will naturally melt in a dramatic fashion. The interval of time between major movements of glaciers may also be very long. In certain areas of Alaska, which were previously covered by glaciers, trees are now growing which are fifty years old. This suggests that the glaciers disappeared long before the anthropogenic increase in carbon dioxide emissions were of any importance. The behaviour of glaciers is not a satisfactory measure of climatic change which is illustrated by the Norwegian glaciers. As a result of heavy snowfall, these glaciers expanded during the warm years between 1988-1993 whereas they lost ice during the record cold year of 1985 (Österholm 1997).

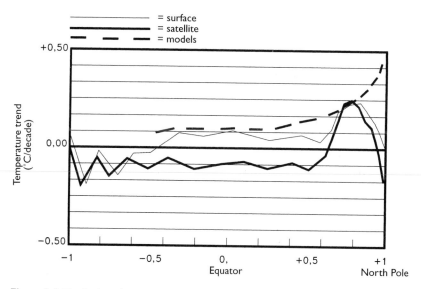

**Figure 5.4** Distribution of temperature trend.

*Arctic areas, Alaska*

The climatic models predict a substantial degree of warming in the polar areas. A substantial proportion of this warming should already have taken place. A new, exceedingly thorough statistical analysis of available temperature data (surface measurements) for the Arctic area for the period 1946-1996 fails to show any sign at all of a significant rise in temperatures (Polyak and North, 1997).

Temperature statistics for two important stations in Alaska, Barrow and Anchorage, are presented below. Several reports in the mass media have

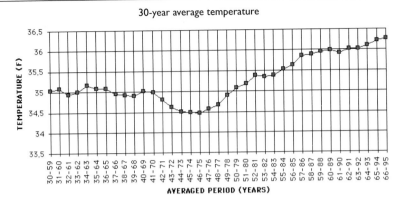

**Figure 5.5** Average temperature for Anchorage 1959-1995 (The Alaska Climate Research Center)

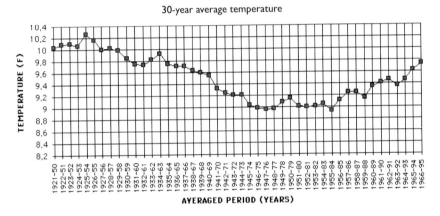

**Figure 5.6** Average temperature for Barrow 1950-1995 (The Alaska Climate Research Center)

alleged that a substantial rise in temperatures has taken place in Alaska and it is naturally of interest to assess whether this is actually the case.

Figure 5.5 presents average temperature data for Anchorage for the years 1959-1995. The weather station located at 61° 14'N has moved several times. Anchorage is a growth region that has a experienced a substantial increase in air traffic. The rise in average temperature that has occurred since 1975 of 0.9°C or 0.45°C/10 years must be at least partly attributable to urban warming. The temperature data from this station cannot therefore be considered to be representative in any discussion of the greenhouse effect. The station is located in the northern part of the industrial belt described above. The warming is concentrated to the coldest winter nights.

The temperature at Barrow is 0.3°C colder than in the 1950s. This is hardly in agreement with most of the predictions from the climatic models which have suggested that the temperature around the seventieth degree of latitude should already have increased substantially as a result of increased emissions of greenhouse gases.

There is nothing unnatural about the climate observations in Alaska. The variations in temperature recorded at the two stations would appear to be a combination of natural variations and urban warming.

### South Pole

Figure 5.7 shows the average surface temperatures for the Amundsen station at the South Pole between 1975 and 1995. There are no indications of any increase in temperatures.

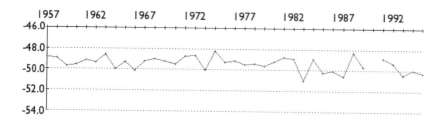

**Figure 5.7.** Average surface temperatures for the Amundsen station at the South Pole between 1975 and 1995 (Daly 1998)

## Final comments and discussion

Seventeen years ago Madden and Ramanathan (1980) and Wigley and Jones (1981) contended that temperature data could not be used to directly establish a anthropogenic causal relationship with the greenhouse effect. Although we have now had access to satellite measurements for almost twenty years and experienced a 17 per cent increase (55ppmv) in the atmosphere's carbon dioxide ratio, the conclusion to be drawn today must be the same. The identification referred to in the IPCC's Summary for Policymakers is indirect and must be called into question.

The ice collapse phenomena in the western Antarctic has nothing to do with climatic conditions (Bentley 1997). The connection made between this phenomena and the carbon dioxide content of the atmosphere is just as absurd as the linkage made with the floods and the deliberately started forest fires in Indonesia in the autumn of 1997.

The temperature data from Siberia has indirectly played a considerable role in the discussion of the greenhouse effect since they provide a substantial con-

75

tribution to global warming as indicated by the surface temperature measurements which could not however be verified by the satellite measurements. In this context, I would wish to raise a warning voice. During the Communist era, heating fuel was distributed in Siberia by means of central edicts and a complicated system of rules and regulations. The supply of fuel available was dependent on how cold it was. For this reason, there may have been a tendency to adjust the temperature readings downwards. After 1990, this system has been abolished in stages and the need to provide exaggerated low temperature figures has gradually disappeared.

## References

Bentley, C.R. (1997), *Science*, 275: 1077-1078.

Christy, J. (1995), USGCPR National Seminar, 20 May 1995.

Christy, J, Mc Nider, R.T. (1994), *Nature* 367: 325.

Daly, J. (1998), *http://www.vision.net.au/~daly* .

Hegerl, G.C. et al. (1997),      *Climate Dynamics* 13: 613-634.

IPCC (1995), Climate Change 1995 - The Science of Climate Change. Cambridge University Press.

Landsea, C.W. *et al.* (1996), *Geophysical Research Letters* 23 (13): 1697-1700.

Lindzen, R. (1994), *Annual Rev. Fluid. Mech* 26: 353-378.

Madden, R.A., Ramanathan, V. (1980), *Science* 209: 763-768.

Polyak, I., North, G. (1997), *J. of Geophysical Research* 102: 1921-1929.

Santer, B. et al. (1996), *Nature* 382: 39.

Tett, S. et al. (1996), *Science* 274: 1170.

Wigley, T.M.L., Jones, P.D. (1981), *Nature*, 292: 205-208.

Österholm, H. (1997), *Hufvudstadsbladet* 17.12.1997.

# Model predictions and future climate changes

# 6.

# Energy use and climate models

*Tor Ragnar Gerholm*

As a physicist, I have no reason to doubt that the IPCC's meteorologists have done their utmost to present a scientific analysis of the issues on which they have been consulted. Nor would I seek to deny that it is meteorologists rather than physicists who are the experts on climate.

However, strictly speaking, it is now a question of climate change rather than climate *per se*. Moreover these changes, according to the IPCC, are attributable to human intervention, namely the emission of greenhouse gases, particularly carbon dioxide, resulting from our use of fossil fuels.

The extent of the feared changes in climate is therefore ultimately bound up with our future use of coal, oil and fossil gas, over the next hundred years or more. These were the issues debated at the Kyoto conference, not meteorology.

In relation to questions of global energy use, meteorologists in the IPCC cannot be said to have any special knowledge.

Moreover, it would not be an exaggeration to state that the IPCC has until now treated the issues of energy use, that are of decisive importance for the Panel's conclusions, in a highly cursory fashion.

The IPCC report '*Climate Change*' (IPCC 1990) was the basis for the convention on climate change adopted at the Rio conference in 1992. Half a page of this 300 page report was devoted to future carbon dioxide emissions. It contained a scenario with the politically emotive title '*Business-as-Usual*'(BAU). This scenario proved subsequently to be of such inadequate quality that it was rejected by the IPCC prior to the Rio conference, a decision which seemed to escape the attention of most of the ten thousand conference delegates.

The *Second Assessment Report* (SAR) published by the IPPC in 1996 provided the scientific basis for the Kyoto conference. No new emission scenarios were presented. Instead reference was made to six scenarios -IS92 a-f - that had been added at the last moment prior to the Rio conference. This seemed like adding the yeast after the dough.

No mention is made in the SAR report of the devastating criticism of these scenarios in the years following their publication.

What is even worse is that these scenarios in the judgement of the IPCC were also considered to be unacceptable. Consequently, they were rejected as well, after the Kyoto conference.

This means that at the current moment, given the lack of acceptable emission scenarios, we have no actual knowledge about what may possibly happen to the climate in the next hundred years. Nor are we likely to be able to acquire such knowledge in the near future. What we do know however is that we could well afford to wait before undertaking expensive corrective measures until we have a sounder basis for such decisions.

In the following, I will present support for the two statements that I made above. In doing so I will restrict myself to the IPCC's own reports and to statements made by leading representatives of the IPCC.

## Scenarios and models

One can hold different views regarding the scientific value of climate models. However one thing is abundantly clear: Climate models are not able *per se* to make any predictions about future levels of carbon dioxide in the atmosphere. The models instead state that if the carbon dioxide content of the atmosphere increases by a certain amount, the surface temperature of the earth will (perhaps) increase by so and so many degrees.

This is hardly a controversial statement - on the contrary. Already in the first IPCC report, *Climate Change* (IPCC 1990), the following was stated in the *Policymakers Summary* (p. xxvii ):

"Uncertainties in the above climate predictions arise from our imperfect knowledge of
- future rates of human-made emissions
- how these things will change the atmospheric concentrations of greenhouse gases
- the response of climate to these changed concentrations

Firstly, it is obvious that the extent to which climate will change depends on the rate at which greenhouse gases (and other gases which affect their concentrations) are emitted. This in turn will be determined by various complex economic and sociological factors. Scenarios of future emissions were generated within IPCC WG111 and are described in the Annex to this Summary."

79

A similar statement is to be found in SAR's Summary for Policymakers (IPCC 1996) where it is said that the IS92 a-f scenarios

"can be used to project atmospheric concentrations of greenhouse gases.... Climate models can *then* ( my italics) be used to develop projections of future climate"

In other words, in order to use the climate model estimates of future changes in climate, it is necessary to proceed from one or several scenarios. Naturally this implies that the climate change estimates cannot be entitled to any greater credibility than can be given to the underlying scenarios.

Nor can they be used to make forecasts. The continuation of the quote above refers to "best estimate" values which is misleading and contrary to the recommendations of those who devised the IPCC scenarios. Scenarios cannot be used in this way.

## Scenarios

In the public debate, a confusion is frequently made between scenarios and forecasts. However there is a fundamental difference between these two concepts. A *forecast* is a scientifically based and empirically tested prediction. On the other hand, a *scenario* is a term used in studies of the future to denote 'a systematic description of a *possible* future situation and of a *conceivable* development from today's situation to the described state of affairs'.

There is no doubt that the authors of the IPCC scenario view their six alternatives, IS92 a-f, in the above light.

The possible situation in the year 2100 represents in all six cases[1]

"a global energy system which continues to be dependent on fossil fuels, notably coal, and which shows only moderate gains in energy efficiency and technological development of non-fossil fuels"

Presumably the general reader does not realise that the IPCC climate projections are covered by the caveat quoted above. There are those who doubt that the world's energy requirements will be entirely or primarily based on coal for the next hundred years. Nor is it considered improbable by many that the efficiency of energy use will continue to improve at more or less the same pace as it has done throughout most of the twentieth century in industrial countries. There are also those who believe that the proportion of energy use which will

---

[1] IPCC Emission scenarios. An update 1992.

80

be met from non-fossil energy sources will continue to increase in the same way as during the past 25 years. Anyone who holds all or some of these views regarding the likely development of global energy use must reject the energy futures described in the quotation above. Consequently they are not be able to accept, as I am not, any of the six IPCC scenarios or any of the statements regarding future climate change on which they are based.

The authors of the IPCC scenarios are anxious to point out that their results should not be viewed as predictions. They state:[2]

> "Scenario outputs are not predictions of the future and should not be used as such; they illustrate the effect of a wide range of economic, demographic and policy assumptions. They are inherently controversial because they reflect different views of the future. The results of scenarios can vary considerably from actual outcome even over short time horizons."

Moreover none of these six scenarios should be considered as more probable than any of the others. There is no "mid-range scenario" or "best estimate" [3]

Furthermore, considering the degree of uncertainty about future emissions, it is recommended to use the full range of IS92 scenarios as input to atmosphere/climate models rather than any single scenario.

However the argument proceeds even further. A sensitivity analysis has been carried out. The results are summarised in the following manner :[4]

> "Comparison of reference and policy scenarios suggests that uncertainties in driving forces of emissions within reference cases are at least as large as the uncertainties between reference and policy cases. This implies that over a time horizon of 50 to 100 years it is difficult to distinguish between the effects of climate policies and the effect of societal/technological changes, irrespective of climate policies."

As will seen below, this modesty is well justified.

## "The mid-range scenario"

What is improperly presented as an "a mid-range scenario", i.e. IS 92a, forms the basis on which the IPCC estimate future increases in average temperatures. (SAR, Summary for Policymakers, p5):

---

[2] 1992 IPCC Supplement. Scientific Assessment of climate Change
[3] IPCC Working group III (WGIII/2nd.DOC, Rev 1) 1994
[4] IPCC Working group III (WGIII/2nd.DOC Rev 1) 1994

For the mid-range IPCC emission scenario, IS92a, assuming the "best esti-mate" value of climate sensitivity[5] and including the effects of future increases in aerosol, models project an increase in global mean surface tem-perature relative to 1990 of about 2°C by 2100. This estimate is approxi-mately one third lower than the "best estimate" in 1900.

The probability of this most probable value depends primarily on the degree to which one believes in the "mid-range" scenario" IS92a.

It is not easy to come to any general conclusion on this question. The IPCC's Second Assessment Report consists of three volumes containing just over one thousand pages. Despite the scope of the document, the reader will search in vain for a description of the underlying assumptions for the reference scenarios.

The essential information  is to be found elsewhere, namely in the 1992 IPCC supplement. The data for IS92a is presented in Table 6.1.[6]

The mid-range scenario assumes an average annual rate of economic growth for the period up to 2100 of 2.3 per cent. This would result in an almost unbelievable improvement in welfare. By the end of the next century, the people of the world would be enjoying a standard of living that exceeds present day conditions in advanced industrial countries.

However there is a reverse side to this coin. In order to achieve these impressive welfare levels, our future societies will consume vast amounts of fossil fuel. By the end of the next century, mankind is going to emit 20.3 bil-lion tons of carbon into the atmosphere, primarily in the form of carbon diox-ide from burning coal. This is in fact a higher carbon dioxide emission per capita than current levels in Sweden. In the year 2100, it will then be a ques-tion of the entire world with a population twice as large as the present. A hun-dred years is a long time. It is hardly possible today to gain any concrete idea of what this best estimate scenario is likely to be.

---

[5] Climate sensitivity refers to the average increase in temperature of the surface of the earth brought about by a doubling of the "pre-industrial" concentration of carbon dioxide in the atmosphere. The latter is taken to be 280 ppmv.
[6] These scenarios have been subsequently updated. However the corrections are small and have no bearing on the argument.

**Table 6.1** Assumptions in IPCC's scenario IS92a.

**Table 6.1a** – general assumptions

| World Population 2100 | 11.3 billion |
|---|---|
| Economic Growth 1990–2100 | 2.3% per year |
| Energy Resources | 12 000 Ej conventional oil |
| | 13 000 Ej natural gas |
| | Solar energy at a cost of $0.075/kWh |
| | 191 Ej biofuels at a cost of $70/barrel of oil |
| | equivalents |
| Other | Reduction of SOx, NOx, and VOC-emissions |
| | according to existing international agree- |
| | ments |
| CFC | Partial application of the Montreal Protocol |
| | Gradual discontinuation of CFC-usage in |
| | developing countries until 2075 |

**Table 6.1b** – rates of change

| | 1990 – 2025 | 1990 – 2100 |
|---|---|---|
| Energy intensity TPER/GDP | -0.8% per year | -1.0% per year |
| Coal intensity C/TPER | -0.4% per year | -2.0% per year |
| Cummulative emissions of fossil carbon, net | 285 GtC | 1,386 GtC |
| Deforestation | 678 million hectares | 1,447 million hectares |
| Carbon emissions from deforestation | 42 GtC | 77 GtC |

**TPER** = Total Primary Requirement of primary energy

**Table 6.1c** – Levels at different points of time

| | 1990 | 2025 | 2100 |
|---|---|---|---|
| $CO_2$-emissions total | 7.4 GtC | 12.2 GtC | 20.3 GtC |
| $CH_4$-emissions | 506 Tg | 659 Tg | 917 Tg |
| $N_2O$-emissions | 12.9 TgN | 15.8 TgN | 17.0 TgN |
| CFC-emissions | 827 kt | 217 kt | 3 kt |
| $SO_x$-emissions | 98 TgS | 141 TgS | 169 TgS |

1 Ej = 42 Gtoe = 42 billion tons oil equivalent
1 GtC = 1 billion tons carbon emitted as carbon dioxide
1 Tg = 1 teragram = $10_{12}$gb = 1 million tons
1 kt = 1,000 tons

The scenario assumes an uninhibited use of the earth's energy resources. It implies that 12 000 EJ (286 Gtoe) oil and 13 000 EJ (310 Gtoe) fossil gas will be made available to meet these demands. According to the IPCC's own assessment SAR, WG II (p 87), present exploitable oil and gas reserves will not be sufficient to meet these demands even if they are supplemented by a probable 50 per cent future increases in reserves. The IPCC has estimated these oil and gas reserves at 8 500 EJ (202 Gtoe)  and 9 200 EJ (219 Gtoe) respectively. The scenario authors consequently rely on technical development that will make "unconventional" oil and gas reserves (asphalt, tar sands and shale) available for the profitable production of energy. However not even this will be enough. The scenario also assumes that practically all available coal reserves of about 600 billion tons will be depleted. This is more than three times the total amount of coal that has already been consumed.

I will leave it to the reader to decide whether or not this energy scenario appears sufficiently probable "to allow it to be used as a basis for predicting future changes in the climate".

However one thing is perfectly clear: It will not occur without a rise in production costs. The prices of fossil fuels will rise relative to those of alternative non-fossil fuels. This is likely to lead to an acceleration of the phasing out of fossil fuels, a development that has already been in process for more than twenty years, despite current low oil and gas prices. As Figure 6.1 illustrates, fossil fuels are gradually losing market share to non-fossil energy alternatives. The world's nuclear power stations are now responsible for a larger energy production, expressed in oil equivalents, than Saudi Arabia, the world's major oil producer.

During the past 150 years, the global energy system has undergone a series of structural transformations. We have moved from firewood to coal and sub-

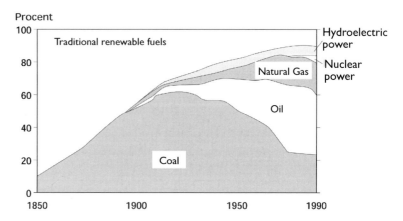

**Figure 6.1** The phasing out of fossil fuels
Source: Global Energy Perspectives to 2050 and Beyond, WEC and IIASA 1995

sequently to oil and gas. During the past thirty years, nuclear and solar power have in different ways increased their small but growing market shares at the expense of fossil fuels. Throughout the whole of this period, carbon intensity has fallen at a fairly even rate (See Figure 6.2). In other words, we use increasingly less carbon per unit of energy input.

This is attributable to two entirely different but mutually supporting factors. Firstly, technical innovation has brought about an increasingly efficient use of energy. Hence, less energy is required per unit of output. Consequently, carbon dioxide emissions will be lower than they would otherwise have been in the absence of technical change. Secondly, the structural changes implicit in Figure 6.1 bring about a gradual increase in the use of less carbon-intense fuels. Table 6.2 provides us with some idea of the importance of this change.

The industrial countries have taken the lead and currently display higher levels of efficiency in energy use and lower ratios of coal and firewood in their fuel mix than developing countries. Given that "the mid-range scenario" assumes a huge increase in welfare which will raise the entire world above the living standards presently prevailing in the industrialised world, it would appear strange to assume that the 150 year trend towards lower carbon intensity will now be broken. According to IS92a, carbon intensity will be almost as high in 2100 as in 1990. The difference is less than 10 per cent.

The reason is quite simply that one assumes that there will be a continued dependence on fossil fuels, particularly coal, and that the rate of technical innovation will be modest with respect to more efficient energy use and to the development of non-fossil energy sources. No arguments are presented in support of these assumptions. In short one assumes what is to be proven.

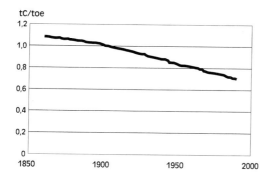

**Figure 6.2** Carbon intensity of the world's primary sources of energy.
Source: Global Energy Perspectives to 2050 and Beyond, WEC and IIASA 1995.

85

**Table 6.2** Proportion of carbon per unit of energy input
for various forms of energy

| Types of energy | GtC/Gtoe | Comment |
|---|---|---|
| firewood | 1.25 | non-sustainable use |
| coal | 1.08 | |
| oil | 0.84 | |
| fossil gas | 0.64 | |
| biomass | near *0 | sustainable use |
| nuclear | near 0 | |

Source: WEC/IIASA 1994 p.42
GtC = billion tons coal
Gtoe = billion tons oil equivalent
* small amounts of carbon dioxide are emitted in primary energy production

If one instead makes the reasonable assumption that carbon intensity will at least continue to decline at the same rate in the future as it has done for the past 150 years, annual carbon dioxide emissions will fall to 14.4 GtC per annum by the end of the next century rather than the scenario figure of 20.3 GtC per annum. This means that the atmospheric concentration of carbon dioxide during the whole of the next century will be below the magic figure of 560 ppmv which denotes a doubling of the "pre-industrial" carbon dioxide content of the atmosphere. Given unchanged assumptions in all other respects, the "most probable" IPCC figure for the global temperature increase will accordingly fall by as much as 25 per cent or to 1.5°C by the end of the next century.

This arithmetic example shows that the IPCC temperature forecasts are actually highly sensitive to changes in assumptions regarding future energy use. The World Energy Council (WEC) and the International Institute for Applied Systems Analysis (IIASA) have in a joint report (WEC/IIASA 1995) reached the same conclusion. In what they consider to be the most realistic of the six scenarios that they develop, they arrive at an estimate of carbon dioxide emissions at the end of the next century, 14 GtC/per annum, that coincides almost exactly with the result of the calculation above. The WEC/IIASA scenario receives the following comments:

> "Overall, even with comparatively modest expectations of technological change, an orderly transition away from fossil use is not only feasible but appears also manageable in terms of energy sector and institutional adjustments extending toward the end of the 21st century."

86

## History repeats itself

The IPCC report *"Climate Change"* (1990) served as the major document for the Rio convention, its conclusions being based on the *Business-as-Usual* scenario. However as was already stated, this scenario had already been rejected prior to the Rio convention.

In a similar fashion, the IPCC report to the Kyoto conference, the Second Assessment Report (1995) along with its updated scenarios, IS92 a-f have been used to predict future changes in climate. These scenarios have also been jettisoned. A British coal expert, Ken Gregory, has been appointed by the IPCC to lead the work on developing new emission scenarios. It is envisaged that this work will be completed by 2001.

In a frank statement, Ken Gregory contended (Gregory 1997):

> "The problems with the existing scenarios arise partly because of the limited time that was available for their development and the consequent lack of input from a wide range of knowledgeable sources. Following an initial public presentation and discussion of ideas, the lead authors found themselves short of time to develop the projection of emissions and were unable to interact with outside reviewers. For most of the period they were working towards developing one main scenario and a variant of it, but added four more in the run up to the plenary at which they were agreed to give a wide range of alternative projections. Projections of sulphur dioxide were also added at the last minute. On top of this, details of the underlying assumptions only became available to reviewers some three months after the scenarios had to be agreed. Thus, it comes as no surprise that when we look back at the scenarios we can find fault. Hopefully we will do better this time."

## Can we wait before undertaking expensive corrective measures?

The answer is undoubtedly *yes*.

In order to support my case, I will present the following argument *a fortiori*. Let us assume that we have already experienced a global temperature increase of 0.5 per cent and that it is solely attributable to anthropogenic emissions of greenhouse gases. We assume furthermore that we can rely on the IPCC climate models and that they provide us with correct predictions of the future climate changes that will result from enhanced concentrations of greenhouse

---

[7] The scenarios IS92 a-f are all below BAU.

gases in the atmosphere. Finally we assume that the carbon dioxide emissions will increase according to the original Business-as-Usual scenario.[7] We ignore the criticism that has been directed at each of these four assumptions and instead ask ourselves, given these assumptions, what the likely consequences would be of a ten years' moratorium on any climate policy measures.

The greenhouse effect will of course be reinforced compared with what otherwise would have been the case if emissions had been held constant. However this increase will not be particularly large - at the most around 7 per cent and by 2100 around 1.5 per cent.

Such a slight increase in the greenhouse effect cannot be demonstrated by scientific methods. Nor will it be of any practical importance. Why will the effect be so limited?

Firstly the difference between the volume of carbon dioxide that would have been added to the atmosphere according to the BAU scenario is only marginally greater than in the stabilisation scenario. Secondly the slight increase in the greenhouse effect is proportional to the logarithm of the carbon dioxide content of the atmosphere. Hence the effect of a marginal increase is less than in the case of proportionality.

The counter argument can be raised that it would be more difficult to introduce restrictions after ten years since we would have become more dependent on fossil fuel than is currently the case.

However this is a question of energy use and economic policy rather than a climatological matter and is in all probability incorrect.

It is likely that greater political support will become available once there is a firm empirical foundation for a manmade greenhouse effect. As developments after both Rio and Kyoto have shown, there is little opportunity at present for any binding agreements on a world-wide stabilisation of carbon dioxide emissions, not least due to the opposition of developing countries who refuse to accept any such restrictions.

On the other hand, carbon dioxide emissions have increased more slowly during the 1990s. In the OECD area, there has been a modest increase of 4 per cent altogether between 1990 and 1995. For the world as a whole, the increase is no more than a total of 3 per cent. In global terms, it should also be noted that carbon dioxide emissions per capita have been constant over the past twenty-five years.

## Conclusions

The above discussion should not be interpreted as a recommendation of total passivity. There is undoubtedly a serious risk that carbon dioxide and other greenhouse gas emissions that result from human activities will have an impact on the earth's climate. It is probable that the effects of this influence will be largely negative.

The threat of changes in global climate must therefore be taken seriously. My argument here is that we both can and should wait before undertaking *expensive* corrective measures.

There is a lot that could be done which does not involve heavy outlays of expenditure. In certain cases, there may even be savings. We should for example ask for a withdrawal of the fairly widespread use of subsidies given to the production or consumption of fossil fuels. We should encourage the introduction of more efficient uses of energy, particularly in the former centrally planned economies and in the third world. Efforts should also be made to eliminate the obstacles to the introduction of non-fossil fuel alternatives.

The development of a regulatory system for joint implementation should also be seen as an urgent priority. Here steps should be made to secure both bi-and multilateral agreements on efficient ways to reduce carbon dioxide emissions. Moreover, we could also start to prepare the international treaties that will be required if the earth's climate is seen to be under serious threat from anthropogenic emissions of greenhouse gases.

Here in Sweden, we could contribute by abandoning our current plans to phase out nuclear energy. Despite a high level of energy consumption, Sweden is able to maintain one of the lowest per capita levels of carbon dioxide emissions in the industrialised world. This is mainly attributable to the country's programme of nuclear energy.

## References

IPCC (1990), *Climate Change - The IPCC Scientific Assessment.* Cambridge University Press.

IPCC (1992), *Climate Change 1992 - The Supplementary Report to the IPCC Scientific Assessment.* Cambridge University Press.

IPCC (1996a), *Climate Change 1995 - The Science of Climate Change.* Cambridge University Press.

IPCC (1996b), *Climate Change 1995 - Scientific-Technical Analysis of Impacts, Adaptions and Mitigation of Climate Change.* Cambridge University Press.

IPCC (1996c), *Climate Change 1995 - The Ecomonic and Social Dimensions of Climate Change.* Cambridge University Press.

WEC/IIASA 1991 (1995), *Global Energy Perspectives to 2050 and Beyond.* N. Nakićenović (Study Director). WEC/IIASA Report.

Gregory K. (1997), *New Scenario Work for Greenhouse Gas Emissions.* Ecoal 2 March 1997.

# 7.

# The carbon dioxide content of the atmosphere

*Jarl Ahlbeck*

This chapter will be concerned with a statistical analysis of data on the carbon dioxide content of the atmosphere and anthropogenic carbon dioxide emissions between 1970 and 1997. With the help of this data, estimates can be made of the rate of net absorption i.e. the exchange between carbon dioxide emissions and the biosphere and the world's oceans. These results are initially presented in a summarised form. The analysis is then developed with the help of mathematical formulae and an explanatory text for those who wish to follow the details of the argument.

Only a small fraction of the $CO_2$ emissions released into the atmosphere actually remain there. This fraction is usually referred to as the airborne fraction. In the latest IPCC report, it was estimated to be 46 per cent.

Carbon dioxide emissions during the period 1970-1996, a period for which there is excellent data, have increased at a linear rate. Our statistical analysis also shows that the carbon dioxide content of the atmosphere displays the same straight line linear trend over this period. Hence the annual increase in the carbon dioxide content of the atmosphere has been approximately constant which implies that the airborne fraction has declined as the carbon dioxide content of the atmosphere has increased. This means that there is a considerable self-damping effect that will moderate the future increase of the atmospheric carbon dioxide concentration.

The analysis may then be checked by estimating the equilibrium value for carbon dioxide content i.e. the value at which the carbon dioxide content of the atmosphere would be constant or as it is stated, pre-industrial. The equilibrium value obtained was 277 ppmv which was exactly the figure obtained by other methods by the IPCC for pre-industrial levels of carbon dioxide.

A projection of these results to 2100 can also be made. I am well aware of the uncertainty surrounding this type of prediction. The results are presented in Figure 7.2. A direct business-as-usual projection of present trends (LEXP) produces a value that is substantially lower than a doubling of pre-industrial levels of carbon dioxide. The average scenario of the IPCC - the IS92a emission scenario - (See Chapter 1, Figure 2 and Table 1.1 for details) produces a slightly higher figure of the future atmospheric concentration but nevertheless much lower than IPCC estimates.

It is also interesting to conduct the mental experiment that if all carbon dioxide emissions were stopped in the year 2000, the level of carbon dioxide in the atmosphere would gradually reach a pre-industrial level in the year 2100.

## Linear increase in the carbon dioxide content of the atmosphere since 1970

A multi-regression analysis of the concentration of carbon dioxide in the atmosphere for the period 1970-1996 (Data from Carbon Dioxide Information Center) indicates that the carbon dioxide content of the atmosphere has increased from 325 to 362 ppmv (ppmv = million parts in relation to volume). The increase is shown to be linear since a quadratic term is eliminated from the analysis as statistically insignificant. The linear term may be written as:

$$C_a = a_o + a_1 \Delta t \qquad (1)$$

where $C_a$ = the carbon dioxide content of the atmosphere (ppmv), $\Delta t = t\text{-}1970$ (year). The regression coefficients are $a_o = 324$ and $a_1 = 1.46$ for the period 1970 -1996,. The standard deviation for $a_1$ is 0.02.

If the concentration of carbon dioxide in the atmosphere is considered to be representative of the entire atmosphere, it is multiplied by a factor 2.123 GtC/ppmv (GtC billion tons of coal) in order to estimate the mass of coal in the atmosphere

$$m_a = d_o + d_1 \Delta t \qquad (2)$$

where $d_o = 688$ and $d_1 = 3.1$. The standard deviation for $d_1$ is 0.04.

## Linear increase in anthropogenic carbon dioxide emissions since 1970

According to Marland et al. (1994) and Jefferson (1997), the anthropogenic emissions of carbon dioxide have increased steadily since 1970. If we assume that the contribution of deforestation is of the order of 1.6 GtC/a which is a figure that is widely used e.g. by the IPCC (1996), the total increase in anthropogenic carbon dioxide emissions has increased from approximately 5.7 GtC/a in 1970 to circa 8.1 GtC/a in 1996. There are two breaks in the curve: the energy crisis of the late 1970s and the changes in eastern Europe in the early 1990s.

A statistical analysis of this data shows that the inclusion of a quadratic term is not statistically significant. The relationship is consequently linear and may be written as

$$F_{em} = b_o + b_1 \Delta t \qquad (3)$$

where $F_{em}$ represents the total anthropogenic carbon dioxide emissions (GtC/a) and $b_o = 5.86$ and $b_1 = 0.085$ are the regression coefficients. The standard deviation for $b_1$ is 0.004.

## Only a part of the emissions have accumulated in the atmosphere

The derivative of Eq. (2) provides us with a measure of the rate at which anthropogenic carbon accumulates in the atmosphere:

$$F_{acc} = \frac{dm_a}{d\Delta t} = d_1 \qquad (4)$$

where $F_{acc}$ is the rate of carbon dioxide accumulation (GtC/a) which has been constant. The emissions that have not been accumulated in the atmosphere have been absorbed by the oceans and the biosphere, the so-called, carbon dioxide sink The rate of net absorption is obtained from the balance:

$$F_s = F_{em} - F_{acc} = b_o - d_1 + b_1 \Delta t \qquad (5)$$

where $F_s$ is the net rate of absorption GtC/a).

Since both equations (2) and (5) contain the term $\Delta t$, time may be eliminated from the equations. In this way, we obtain an expression that describes how the carbon dioxide content of the atmosphere or the atmosphere's total coal content affects the rate of absorption. According to well established laws in physical chemistry and biology, the rate of absorption in both fluid and organic form is dependent on the partial pressure in the gas phase. Due to the complexity of the system, it is difficult to determine the structure of this relationship.

However our statistical analysis yields a simple linear relationship

$$F_s = g_0 + g_1 \, m_a \qquad\qquad (6)$$

$$g_0 = b_0 - d_1 - \frac{b_1 d_0}{d_1} = -16.1 \qquad\qquad (7)$$

$$\text{and } g_1 = \frac{b_1}{d_1} = 0.0274 \qquad\qquad (8)$$

In order to test whether equation (6) has any basis in reality, let the rate of absorption be equal to zero and estimate the amount of atmospheric carbon that is required to produce a net absorption capacity equal to zero and equilibrium in the system. We obtain $16.1/0.0274 = 587.6$ GtC.

*By dividing 587.6 GtC by the conversion factor 2.123 GtC/ppmv, we obtain 277 ppmv as the equilibrium concentration of carbon dioxide.* Hence the statistical analysis of the best data available for the period 1970-1996 would appear to indicate that a pre-industrial level of around 280 ppmv would be a reasonable assumption. Moreover this calculation verifies the fact that the net rate of absorption cannot be far from a linear function of the partial pressure of carbon dioxide. Even when we use substantially different values than 1.6GtC/a for the contribution from deforestation, we obtain more or less the same results.

## Relative sinks and the airborne fraction.

The relative sink, $f_s = F_s / F_{em}$ can now be estimated from the regression coefficients We thereby obtain

$$F_s = 1 - \frac{d_1}{b_0 + b_1 \Delta t} \qquad\qquad (9)$$

Inserting numerical values, we obtain 47 per cent for 1970 and 61.5 per cent for 1996. Using 95 per cent confidence intervals, the range within which the actual value is likely to fall with 95 per cent certainty is 43 -51 per cent and 59-64 per cent respectively. Hence the increasing trend of the relative sink is statistically significant.

Analogously, the airborne fraction can be estimate as $1-F_s$. Figure 7.1 shows how it has declined from 53 per cent to 38.5 per cent between 1970 and 1996. A 95 per cent confidence interval is included in the diagram.

What would it mean if we assumed, like the IPCC, that the relative sink and the airborne fraction were constant?

93

Since the emissions have increased in linear fashion, a general mass balance may be written as follows:

$$\frac{dm_a}{d\Delta t} = F_{em} - F_s = F_{em}(1-f_s) = (b_0 + b_1\Delta t)(1-f_s). \quad (10)$$

The separation of the variables and the integration gives the following expression:

$$\int_{m_a(1970)}^{m_a(1970 + \Delta t)} dm_a = b_0 \int_0^{\Delta t} (1-f_s)d\Delta t + b_1 \int_0^{\Delta t} (1-f_s)\Delta t d\Delta t \quad (11)$$

If we insert a constant value for the relative sink, we obtain :

$$m_a(1970 + \Delta t) = m_a(1970) + (1-f_s)b_0\Delta t + (1-f_s)b_1 \frac{\Delta t^2}{2} \quad (12)$$

If the relative sink was to be constant, as implied by the IPCC analyses, the carbon dioxide curve for the years 1970-1996 would have taken the form of a parabolic segment with a positive second degree coefficient. However the second degree coefficient produced by the regression analysis is statistically insignificant. Hence the available data clearly shows that the relative sink ought to have increased and that the airborne fraction should have declined.

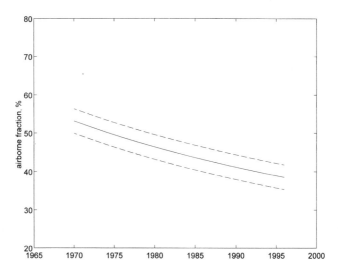

Figure 7.1 Atmospheric share (&) for the years 1970-1997 at 95 per cent confidence intervals.

94

In addition, a possible theoretical basis for a constant relative sink should be subject to close scrutiny since it would imply that the rate of absorption (in GtC/a) would be dependent on the derivative of time for the partial pressure of carbon dioxide instead of the actual partial pressure.

## The calculations for the next century

If we wish to estimate the carbon dioxide content of the atmosphere for the next century solely on the basis of the statistical data for 1970- 19967and without the use of physics-based models, we will have to assume that the relationship between the rate of absorption and the atmosphere's partial pressure of carbon dioxide will be constant between 1970 and 1996. This would appear to be a reasonable assumption in the light of the accuracy of our backward extrapolation to a pre-industrial level.

We must also make an intelligent guess about the trend of anthropogenic emissions

Let us define a new time interval, $\Delta t$, which is zero for the year 2000 and 100 for the year 2100. We obtain the following equation for the anthropogenic carbon dioxide emissions:

$$F_{em} = F_{em,0} + h\Delta t \qquad (13)$$

where $F_{em,0}$ represents the total anthropogenic carbon dioxide emissions in the year 2000 or 8.41 GtC/a. The coefficient h may be given a negative value for a reduced level of emissions, zero for unchanged levels of emissions and a positive value for increasing levels of emissions.

The mass balance may then be expressed as follows:

$$\frac{dm_a}{d\Delta t} + F_s = F_{em} \qquad (14)$$

$$\frac{dm_a}{d\Delta t} + g_1 m_a = h\Delta t + F_{em,0} - g_0 \qquad (15)$$

The solution of the differential equation and the insertion of the limit condition $m_a = m_{ao}$ (ca. 781 GtC or 368 ppmv) for $\Delta t = 0$ (year 2000) gives us the following equation for the carbon content of the atmosphere as a function of time:

$$m_a(\Delta t) = \frac{h\Delta t}{g_1} + m_{a,0}\exp(-g_1\Delta t) + \left[\frac{F_{em,0}}{g_1} - \frac{h}{g_1} - \frac{g_0}{g_1}\right].[1 - \exp(-g_1\Delta t)] \qquad (16)$$

95

Dividing by the factor 2.123 GtC/ppmv expresses the carbon dioxide content of the atmosphere in ppmv.

The model is entirely based on the statistical data for the years 1970 -1996. No other relationships are used, not even the available knowledge that the pre-industrial level was roughly 280 ppmv. Despite these restrictions, the model succeeds in estimating the pre-industrial level.

### Scenario IS92 a

In this IPCC scenario, $b = 0.1146$. The carbon dioxide concentration according to this scenario is shown in Figure 7.2 (IS92a). The result obtained for the year 2100 is approximately 549 ppmv. which is far below the figure of 705(!) ppmv. quoted by the IPCC (1996) p.23, Fig.5 (b).

### Continuing linear increase of carbon dioxide emissions

If the anthropogenic carbon dioxide emissions continue to increase as during the years 1970-1996, the model will naturally show that the carbon dioxide content of the atmosphere will continue to increase at the same rate as previously. By 2100, the concentration will be circa 514 ppm.
$b = b_1 = 0\ 0.085$. See figure 7.2 (LEXP).

As was the case with IS $_{92}$a, the LEXP appears unrealistic when the use of fossil fuel exceeds 15GtC/a in 2100. This would mean that over 1000 GtC of the remaining reserves from 2000 would have been used. The total reserves (known and hypothetical) would appear to be circa 1500 GtC (Treutlin and Ahlbeck 1997). The rate of deforestation would also be kept at a level of 1.6 GtC/a.

### Zero increase in emissions after 2000

Nor does this projection appear realistic. The best guess would appear to be somewhere between LEXP and zero increase. Setting $b = 0$, we obtain 416 ppmv for 210, see Fig.7.2.

### Stop all emissions

In order to test the model, we conduct the mental experiment that all emissions are stopped in 2000 by setting Fem,0 and h equal to zero . The carbon

96

dioxide content of the atmosphere will then gradually decline to pre-industrial levels, see Fig. 7.2.

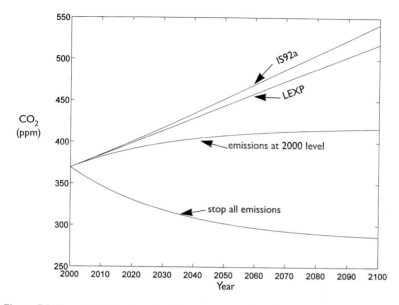

**Figure 7.2** The estimated carbon dioxide concentration (ppmv) in the atmosphere during the next hundred years.

## References

IPCC (1996), *Climate Change 1995 - The Science of Climatic Change*, Cambridge University Press.

Jefferson, M. (1997), Carbon dioxide emissions 1990-1996, *WEC Journal*, July 1997.

Marland, G., Andres, R.J., Boden, T.A. (1994), *Global, regional and national $CO_2$ emissions (+update)*, Carbon Dioxide Information Center.

Treutlein, S., Ahlbeck, J. (1997), *Fossil Fuel Resources*, Abo Akademi University, Process Design Laboratory, report 97-15-B.

# 8.

# The Greenhouse Effect and its Problems

Richard S. Lindzen
*Dept. of Earth, Atmospheric, and Planetary Sciences*
*Massachusetts Institute of Technology*

## *Introduction*

The phrase greenhouse effect in the context of climate refers loosely to the assertion that since the earth's atmosphere is relatively transparent to solar radiation, and relatively opaque in the infrared, due to the presence of substances which absorb in the infrared (the so called greenhouse gases: mainly water vapor and condensed water in the form of clouds, but also carbon dioxide, methane, nitrous oxide, and ozone), the surface temperature must be warmer than it would be in the absence of these greenhouse gases in order for surface cooling to balance incoming solar radiation. As evidence for this view, it is noted that the earth's surface is, indeed warmer than it would be in the absence of an atmosphere. It is usually further claimed that an implication of this picture is that increasing greenhouse gas concentrations will increase surface temperatures. Figure 1 in IPCC 90 illustrates this view – focusing only on radiative heat exchange and inaccurately portraying even this case. This line of reasoning has a long history, dating back to Fourier and Arrhenius (1896) in the 19th Century.

There are, however, many reasons to question the hypothesis. Some basic arguments run as follows[1]:

1. The basic greenhouse process is not simple. In particular, it is not simply a matter of the gases which absorb heat radiation (greenhouse gases) keeping

---

[1] A general view to the physics of climate may be found in Lindzen, 1994.

the earth warm. If it were, the natural greenhouse would be about 4 times more effective than it actually is. In reality, the surface of the earth is cooled by evaporation and motion systems which bodily carry heat both upwards and polewards, thus bypassing much of the atmosphere's greenhouse absorption. The actual greenhouse effect depends on these motions as well as the greenhouse gas concentrations above the levels where motions deposit heat and the details of the temperature distribution at these levels. All of these are matters of significant basic uncertainty, and involve errors in model behavior so large as to be discerned even in the uncertain data.

2. The most important greenhouse gas in the atmosphere is water vapor, and percentage changes in this gas are comparably important at all levels of the atmosphere (at least below 16 km) despite the fact that the concentration of water vapor is thousands of times less at 16 km than at the surface. Roughly speaking, changes in relative humidity on the order of 1.3-4% are equivalent to the effect of doubling $CO_2$. Our measurement uncertainty for trends in water vapor is in excess of 10%, and once again, model errors are known to substantially exceed measurement errors in a very systematic way. It should be added that the radiative effect of water vapor is nonlinear, and the effect of small changes in dry regions will matter much more to the radiative balance than changes in moist regions. It is the dry regions that have been most poorly measured.

3. The direct impact of doubling $CO_2$ on the earth's temperature is rather small: on the order of 0.3C (Lindzen, 1995). Larger predictions depend on positive feedbacks, primarily from upper atmosphere temperature and from water vapor, acting in such a manner as to greatly magnify the effect of $CO_2$. Both these factors arise from models with errors in these factors, the importance of which is likely to greatly exceed the effect of doubling $CO_2$.

There is very little argument about the above points. They are, for the most part, textbook material, showing that there are errors and uncertainties in physical processes central to model predictions that are an order of magnitude greater than the climate forcing due to a putative doubling of $CO_2$. There is, nonetheless, argument over whether the above points mean that the predicted *significant* response to increased $CO_2$ is without meaningful basis. Here there is disagreement. Major users and developers of large models frequently defend model results regardless of the above. Theoreticians and data analysts are commonly more skeptical. The word, *significant*, should be emphasized. Global mean temperatures fluctuate by 0.25C and more without anyone particularly noticing. It seems most peculiar that such disagreements should be described in terms of contrarians and consensi. In order to understand this, one must turn to a major source of semantic confusion: namely the difference between a natural consensus arising in a field and a forged consensus. It should be added that there is a substantial body of both theoretical and observational analysis that strongly suggests that the models have exaggerated the impact of increasing $CO_2$. However, for present purposes it suffices to note that there is neither an observational basis for concerns nor a credible theoretical basis. Support for the popularly stated scenarios are, at this point, little more than statements of belief rather than science.

The consensus concerning the behavior of the observed globally averaged temperature is pretty much a natural consensus – it has increased about $0.45\pm0.15°C$ since the late 19[th] Century. So too is the consensus over the increase in $CO_2$: it appears to have increased from 280ppmv in 1800 to 360ppmv today. The consensus concerning the model response to increasing $CO_2$, however, is more a forged consensus. Boehmer-Christiansen (1994) describes the issue.

The purpose of the present article is to examine the greenhouse hypothesis more carefully, and to present a more precise formulation in order to illustrate the underlying complexity of the hypothesis, and to examine real and potential weaknesses. Few topics have suffered more from over-simplification. In order to avoid this within the length limitations of this article, we make extensive use of references. In order to facilitate the use of these references, we cite specific figures.

## Thermal equilibrium for an earth without an atmosphere.

The sun behaves approximately like a black body of radius, $r_s=6.599\cdot10^5$ km, at a temperature of $T_s=5783°K$. The radiative flux at the sun's surface is given by the expression $\sigma T_s^4$, where $\sigma$ is the Stefan-Boltzmann Constant ($5.67032\cdot10^{-8}$ Wm$^{-2}$K$^{-4}$). Flux refers to radiation per unit area. Thus, at the earth's distance from the sun, $r_{es}=1.4960\cdot10^8$ km this flux is reduced by the factor $(r_s/r_{es})^2$. The earth's disk has a cross-section, $a_{cs}=\pi r_e^2$, where $r_e$ is the earth's radius ($6.378388\cdot10^3$ km), and thus intercepts $a_{cs}\sigma T_s^4 (r_s/r_{es})^2$ radiation from the sun. In order to balance this intercepted radiation, the earth would warm to a temperature $T_e$, where $\sigma T_e^4 4\pi r_e^2 = a_{cs}\sigma T_s^4 (r_s/r_{es})^2$. This leads to a solution, $T_e=272°K$, which is surprisingly close to the current mean temperature of the earth, 288°K. Somewhat inconsistently, it is generally noted that clouds (which require the presence of an atmosphere) and other features of the earth reflect 31% of the incident radiation. Taking account of this reduces $T_e$ to 255°K.

100

## The radiative role of the atmosphere.

The spectral distribution of radiation of a black body is given by the Planck distribution which is given by

$$B\lambda(\theta) = \frac{2hc^2\lambda^{-5}}{e^{hc/k\lambda\theta}-1}$$

where h is the Planck constant ($6.626176 \cdot 10^{-34}$ Joule second), c is the speed of light ($2.997924580 \cdot 10^5$ km s$^{-1}$), $\lambda$ is the wavelength, and $\theta$ is the absolute temperature in °K. This will differ for bodies at $T_s=5783$°K and $T_e=255$°K. The two distributions are illustrated in Figure 1a taken from Goody and Yung (1989). The spectra are well separated. Figures 1b and 1c from the same reference show the atmospheric absorption spectrum for the total atmosphere above the ground and for the atmosphere above 11 km. The latter height corresponds to a characteristic height of the tropopause. They show that there is relatively little absorption of visible light although gases like ozone, oxygen and nitrogen do absorb ultraviolet radiation primarily in the upper atmosphere. On the other hand the terrestrial radiation is primarily in the infrared where there is extensive absorption by water vapor with other gases like carbon dioxide, ozone and methane contributing where water vapor absorption is weaker.

If the atmosphere cooled only by radiation, then each layer of the atmosphere would both absorb infrared radiation, and emit radiation at that layer's temperature. Thus, for the surface to cool, it would have to be warmer than the air above, and similarly, each layer would have to be warmer than the air immediately above it. Temperature would decrease with height until the opacity of the air above was sufficiently small that radiation could begin to escape directly to space. In point of fact, the radiation to space is characterized by the atmospheric temperature at about one optical depth into the atmosphere (from the top). (An optical depth of 1 corresponds to the path over which radiation is attenuated by $(1/e) \approx 0.368$.) Such a situation is illustrated in Figure 9.10 in Goody and Yung (1989). Such radiative equilibrium leads to surface temperatures of about 350°K which are very much warmer than what is observed. It should be noted that temperatures begin to increase with height above the tropopause because of the direct absorption of ultraviolet by ozone in the stratosphere and mesosphere. In reality the position at which optical depth = 1 depends on wavelength, but we shall ignore this for purposes of simplification.

As was noted long ago by Emden (1913), radiative equilibrium profiles are intrinsically impossible since they lead to such large decreases in temperature with height as to render the atmosphere unstable with respect to buoyant convection.

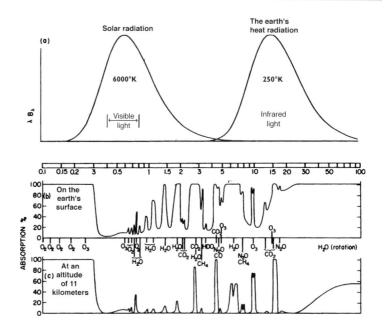

**Figure 8.1.** Atmospheric absorptions. (a) Black-body curves for solar emissions around 6000K and terrestrial emissions around 250K. (b) Atmospheric absorption spectrum for a solar beam reaching ground level. (c) The same for a beam reaching the temperate tropopause. The axes are chosen so that areas in (a) are proportional to radiant energy. Integrated over the earth's surface and over all solid angles, the solar and terrestrial fluxes are equal to each other; consequently the two black-body curves are drawn with equal areas. Conditions are typical of mid-latitudes and for a solar elevation of 40° or for a diffuse stream of terrestrial radiation. From Goody and Yung, 1989.

## Convective adjustment.

In response to the above described problem, a simple fix was developed. The vertical temperature gradient in the troposphere was simply set to its *observed* mean value rather than any particular value one could associate with convection *per se*. Despite this, the procedure is still referred to as *convective adjustment*. The result of this fix is shown in in Möller and Manabe, 1961 (reproduced as Figure 6 in Lindzen, 1994). The surface temperature they obtain is now close to its observed average value of 288K. Of course, this is hardly a prediction. The vertical temperature profile was specifically chosen to match observations, while the latitude and associated solar insolation were chosen to replicate the observed mean temperature. What this simple result does do, however, is to highlight the fact that the surface of the earth, and for that matter the lower layers of the atmosphere, *do not cool primarily by radiation.* Rather, atmospheric motions bodily carry heat upwards, and in a properly 3-dimensional world, polewards (Figure 7, Lindzen, 1994). A more accurate, though still one dimensional picture, is shown in Figure 1.3 in IPCC95 which shows that radiative cooling from the surface is largely canceled by infrared down-welling from the atmosphere. The largest source of net cooling is evaporation, and sensible heat transport is also important. (Sensible heat refers to heat per se as opposed to evaporation which is referred to as transport of latent heat since the heat is only realized when the vapor condenses.) The heat from the last two processes is carried into the atmosphere by both convective towers which carry heat upwards, and the large scale circulation which carries heat both upwards and polewards. The complexity of these motion systems makes it evident that 'convective adjustment' is both a gross oversimplification and a misnomer. Implicit in 'convective adjustment' is a sequence wherein radiation produces a convectively unstable profile which then breaks down convectively. If, however, one allows for a horizontal as well as a vertical dimension, then the horizontal variations in solar insolation automatically give rise to motion systems which lead to stable stratification which, in turn, potentially eliminates the need for simple buoyant convection. The exception is primarily in the tropics where the presence of moisture and its state changes lead to conditional instability which renders the system convectively unstable despite the fact that it would not be unstable in the absence of moisture. Nevertheless, the use of convective adjustment has, as we shall explain in the following section, important implications for the calculated response to increased greenhouse gases.

## The role of vertical dynamic coupling in the response to perturbed greenhouse forcing.

Because of the presence of greenhouse gases, the emission temperature of the earth corresponds to some upper level temperature at approximately one optical depth into the atmosphere rather than the surface temperature. We shall call this level, $z_e$. If one estimates the emission level from a typical General Circulation Model (GCM), one finds that the level varies with location but is typically near 500 mb (or 5.5 km). The particular level depends on the model water vapor distribution which is a matter of controversy. It is useful to consider the role of perturbed greenhouse gas concentrations from the perspective of emission levels. In Figure 8.2 we see the situation for an unperturbed atmosphere. Figure 8.3 shows the situation when one increases the amount of greenhouse gas: obviously, by increasing the total amount of infrared absorbing (or greenhouse) gas, one has elevated the level at which optical depth 1 occurs, and since temperature decreases with height at such levels, the emission temperature, $T_e$, is now reduced (for a doubling of $CO_2$, the level is

103

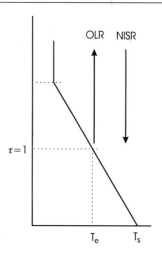

**Figure 8.2.** Radiative balance at the top of the atmosphere. See text for details.

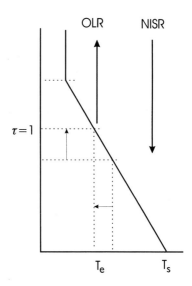

**Figure 8.3.** The imbalance that results from adding to the greenhouse gas level. See text for details.

moved upwards by about 50m). Thus, the outgoing longwave radiation no longer balances the net incoming solar radiation. In order to restore equilibrium, the temperature at the new $z_e$ must rise to the original emission temperature. The relation of the temperature change at the new $z_e$ to temperature change at the surface is dependent on the dynamic coupling of the two levels. Thus, in Figure 8.4, we illustrate three different changes in temperature, each of which allows equilibration with space. In Figure 8.4a we simply hold the

104

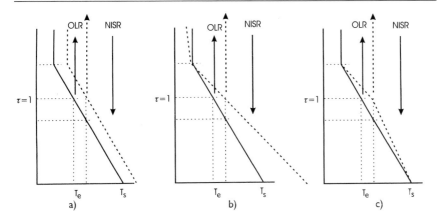

**Figure 8.4.** This figure shows some of the ways in which the imbalance in Fig. 8.3 can be restored. See text for details.

temperature profile fixed while uniformly increasing temperature everywhere. As noted by Lindzen (1995), this approach leads to a very small temperature increase (0.3°K) associated with a doubling of $CO_2$. A problem with this approach is that the stratosphere, which is in approximate radiative equilibrium, must cool when $CO_2$ is increased. In Figure 8.4b, we allow for this, altering the tropospheric lapse rate in order to maintain continuity with the stratosphere. This leads to a much larger surface response.[2] Of course, one could always achieve equilibrium by changing the temperature at $z_e$ without changing the surface temperature at all. This is illustrated in Figure 8.2c. Such a situation might account for the possibility of the earth responding differently to changes in solar output (which directly influence the surface) than to changes in greenhouse gases. In summary, the actual change expected at the surface depends critically on the nature of the dynamic vertical coupling in the troposphere. The use of convective adjustment assumes the coupling is rigid.

## Vertical coupling in nature and models.

That there is an important measure of vertical independence in the thermal structure of the atmosphere is well known. Figure 1 in Lee and Mak (1994) demonstrates this very clearly. Patterns of variance in static stability are clearly different in the 900-700mb layer near the surface from what is found in the upper layer (500-300mb). This is also true in models. However, as Sun and

---

[2] If one insists on allowing both stratospheric cooling, and maintaining a constant unchanged lapse rate, then the response is more complicated, leading to a further elevation in ze and a larger response at the surface (typically on the order of 1.2°K).

Held (1996) demonstrate, there is greater vertical coherence in models than is observed – especially in certain regions. That is to say, in models temperature variations at upper levels in the atmosphere are more tightly coupled to temperature variations at the surface than is found in nature. Relatedly, Held and Suarez (1974) found that OLR correlates far better with temperature at 500mb than with temperature at the surface. This would, of course, be impossible if temperatures at the surface and at 500mb (approximately 6km above the surface) followed each other exactly. As we have shown above, excessive vertical coherence can also lead to excessive sensitivity to perturbations in greenhouse gases.

## Feedbacks

By feedbacks, we generally refer to processes which are affected by surface temperature in such a way as to alter the response of the surface to some initial forcing. Although it is rarely referred to as a feedback, excessive vertical coherence in the temperature field may well constitute the largest positive feedback in current models. Note that such processes may be peculiar to models rather than to nature – in which case the model response may be spurious. Most commonly, the term feedbacks refers to processes which alter either the net incoming solar radiation or the amount of infrared absorbing gas. Thus changes in cloud and snow cover will change the former, while changes in cloud cover and water vapor will change the latter. In this connection, it is important to recall that the most important greenhouse gas in the atmosphere is water vapor. A measure of the importance of water vapor is given in Figure 1 of Lindzen (1997) where it is shown that changes in relative humidity of 5% will (all other things kept constant) alter OLR by from 2-5Wm$^{-2}$ depending on the relative humidity being perturbed with greater responses being associated with smaller unperturbed relative humidity. For purposes of comparison, doubling $CO_2$ decreases OLR by 4Wm$^{-2}$. It should be noted that water vapor at $z_e$ is very variable and measurements of relative humidity are generally uncertain to within at least 10%. Over most of the earth (especially in the tropics), air is subsiding at $z_e$ and frequently originates from thousands of kilometers away (Sun and Lindzen, 1993). This should be contrasted with the air near the surface. The air in the lowest 2 km forms a turbulent boundary layer where the air is well mixed and tightly coupled to the surface; here, warming is indeed associated with increased water vapor. However, the air above the boundary layer has an entirely different origin. Nonetheless, in current GCMs water vapor is tightly coupled in the vertical (i.e., variations in water vapor above the boundary layer closely follow variations in the boundary layer), so that warming inevitably leads in these models to an increase in humidity at all

106

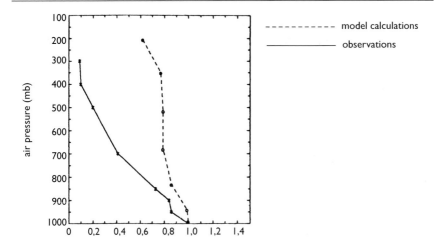

**Figure 8.5.** Vertical structure of the correlations between the variations of specific humidity, qa, at all levels with those at the lowest level. From Sun and Held, 1996.

levels and a positive feedback which turns out to be essential to all predictions of significant warming. However, as Sun and Held (1996) show in their Figure 5, such tight coupling is far from what is observed. This is extremely important for predicted responses to increased anthropogenic greenhouse gases. That there is substantial disagreement between model and observed water vapor is shown in Spencer and Braswell (1997), Bates and Jackson (1997), and Schmetz and van de Berg (1994). Recall that the characteristic discrepancies of 20% in relative humidity are equivalent to discrepancies of 10-20Wm$^{-2}$ in flux. The common assertion that a warmer world will involve more evaporation and hence greater humidity is patently false. Evaporation is balanced by precipitation. Thus, in principle, all increase in evaporation could go directly into precipitation without altering the humidity of the air. The moisturization of the air depends in part on the efficiency of precipitation, with more efficiency associated with less moisturization. In general, precipitation efficiency increases with temperature (Rogers and Yau, 1989).

An approximate expression for the response in terms of feedbacks is

$$\text{radiative effect with feedbacks} = \frac{\text{radiative effect without feedbacks}}{1 - \varepsilon \, f_i}$$

where the $f_i$'s are the various feedback factors. Characteristic feedback factors estimated for GCMs are 0.41 for water vapor, 0.2 for clouds, and 0.1 for snow-ice albedo. The water feedback is clearly the most important, and without it no model will produce a large response to doubled $CO_2$. All the feedbacks are highly uncertain even with respect to sign. The water vapor feedback was first

107

described explicitly by Manabe and Weatherald (1967) who posited a fixed relative humidity so that warming would be accompanied by increased specific humidity everywhere. Current atmospheric observations offer no support for the assumption of fixed relative humidity at levels above the surface boundary layer.

## Further complexity

The above discussion focuses primarily on a simple one dimensional picture of the world. Even this picture is substantially more complex and subtle than one might suppose. However, the real world is not one dimensional. As Spencer and Braswell (1997) show, water vapor varies dramatically with horizontal position, with very dry regions adjacent to extremely moist regions (see also Figure 3 in Lindzen, 1997). Most radiative cooling comes from dry regions which have low, warm emission levels. In practice, the actual greenhouse effect depends significantly on the relative areas of moist and dry regions rather than the areal mean humidity. The same applies to the water vapor feedback.

## Conclusion

We have seen that a proper description of the greenhouse effect does not predict that surface warming must inevitably result in increasing levels of greenhouse gases leading to surface warming. Moreover, we see that model predictions of significant warming are highly dependent on feedbacks in the models whose physical basis is tenuous at best. One may reasonably conclude that current GCMs are inadequate for the purpose of convincingly determining whether the small changes in radiative fluxes at the top of the atmosphere associated with an increase in $CO_2$ are capable of producing significant climate change. However, we may not be dependent on uncertain models in order to ascertain climate sensitivity. Observations can potentially directly and indirectly be used to evaluate climate sensitivity to forcing of the sort produced by increasing $CO_2$ even without improved GCMs. The observations needed for direct assessment are, indeed, observations that we are currently capable of making, and it is possible that the necessary observations may already be in hand, though the accuracy requirements may be greater than current data provides. Still, the importance of the question suggests that such avenues be adequately explored. Since the feedbacks involved in climate sensitivity are atmospheric, they are associated with short time scales. Oceanic delays are irrelevant since observed surface temperatures are forcing the flux changes we are concerned with. The needed length of record must be determined empirically. This is discussed in greater detail in Lindzen (1997). Indirect estimates, based on response to volcanos, suggest sensitivity may be as small as 0.3-0.5C for a doubling of $CO_2$ which is well within the range of natural variability (Lindzen and Giannitsis, 1998). This is not to suggest that such change cannot be detected; rather, it is a statement that the anticipated change is well within the range of what the earth regularly deals with.

# References

Arrhenius, S. (1896) On the influence of carbonic acid in the air upon the temperature of the ground. *Phil. Mag.*, **41**, 237-276.

Bates, J.J. and D.L. Jackson (1997) A comparison of water vapor observations with AMIP I simulations. *J. Geophys. Res.*, **102**, 21,837-21,852.

Boehmer-Christiansen, S.A., 1994. A scientific agenda for climate policy? *Nature*, **372**, 400402.

Emden, R. (1913) Über Strahlungsgleichgewicht und atmosphärische Strahlung. *Sitz. Bayerische Akad. Wiss.*, Math.-Phys. Klasse, **55**.

Goody, R.M. and Y.L. Yung (1989) *Atmospheric Radiation, Theoretical Basis*. Second Edition Oxford University Press, Oxford, 519pp.

Held, I.M. and M. Suarez

IPCC 90 (1990) *Climate Change: The science of climate change*. Cambridge University Press, Cambridge, UK, 365pp.

IPCC 95 (1996) *Climate change 1995: The science of climate change*. Cambridge University Press, Cambridge, UK, 572pp.

Lee, W.-J., Mak, M., 1994. Observed variability in the large-scale static stability. *J. Atmos. Sci.*, **51**, 2137-2144.

Lindzen, R.S. (1994) Climate dynamics and global change. *Ann. Rev. Fl. Mech.*, **26**, 353-378.

Lindzen, R.S. (1995) How cold would we get under $CO_2$-less sky? *Phys. Today*, **48**, 78-80.

Lindzen, R.S. (1997) Can increasing atmospheric $CO_2$ affect global climate? *Proc. Natl..Acad. Sci. USA*, **94**, 8335-8342.

Lindzen, R.S. and C. Giannitsis (1998) The response to volcanos singly and in sequence as a test of climate models. *J. Geophys. Res.* in press.

Manabe, S. and R.T. Weatherald (1967) Thermal equilibrium of the atmosphere with a given distribution of relative humidity. *J. Atmos. Sci.*, **24**, 241.

Möller, F. and S. Manabe (1961) Über das Strahlungsgleichgewicht der Atmosphäre. *Z. Meteorol.*, **15**, 3.

Rogers, R.R. and M.K. Yau, 1989: *A Short Course in Cloud Physics*. Pergamon Press, London, 293 pp.

Schmetz, J., van de Berg, L., 1994. Upper tropospheric humidity observations from

Meteosat compared with short-term forecast fields. *Geophys. Res. Letters*, **21**, 573-576.

Spencer, R.W. and W.D. Braswell (1997) How dry is the tropical free troposphere? Implications for global warming theory. *Bull. Amer. Met. Soc.*, **78**, 1097-1106.

Held, I.M. and M.J. Suarez (1974) Simple albedo feedback models of icecaps. *Tellus*, 26, 613-628

Sun, D.-Z. and I.M. Held (1996) A comparison of modeled and observed relationships between variations of water vapor and temperature on the interannual time scale. *J. Clim.*, **9**, 665-675.

Sun, D-Z. and R.S. Lindzen (1993) Distribution of tropical tropospheric water vapor. *J. Atmos. Sci.*, **50**, 1643-1660.

# 9.

# The role of the oceans in the climatic system

*Gösta Walin*

Almost a hundred years ago, Svante Arrenhius put forward the idea that the mining and burning of fossil fuels would lead to a growing concentration of carbon dioxide in the atmosphere. He estimated that a doubling of the amount of carbon dioxide in the atmosphere would raise the temperature of the surface of the earth by 6°C. This estimate is remarkably close to those under discussion today (in recent years, they have been cut to 2°C to 3°C ).

Arrenhius was aware of the capacity of carbon dioxide to absorb and emit long wave radiation. However he didn't have access to contemporary computerised techniques which lend status and credibility to estimates of future climate change.

However what is remarkable is that our estimates of global warming are just as uncertain as those made in Arrenhius' day. There are many reasons for this situation. Several of the most basic problems that will be discussed in this chapter are to be found in our oceans.

The composition of the atmosphere is largely determined by processes occurring in the oceans. Above all oxygen and carbon dioxide are continuously being replaced by chemical and biochemical processes in the oceans. For example, a doubling of the amount of carbon dioxide in the atmosphere will be reduced to only an 8 per cent increase once the atmosphere has been subjected to these oceanic processes. However the oceans operate slowly. It may take up to 1,000 years to complete these processes. All of the mechanisms involved are not yet fully understood and it is conceivable that they may operate more rapidly.

# The lower boundary of the atmosphere

The oceans cover 70 per cent of the surface of the planet. For the atmosphere, the surface of the ocean represents a "boundary condition". At this interface, we encounter in a concentrated form, the difficulties that confront climate research.

At the surface of the ocean, the atmosphere is influenced by the distribution of temperature, salinity, ice cover and by flows of heat, evaporation etc. Together with the sun's radiation and the boundary constraints over the continents, the state of the atmosphere is determined largely by these conditions at the surface of the ocean.

A short cut that can be used to describe the behaviour of the atmosphere is to use the observed and estimated values of these conditions at the surface of the ocean. By testing different distributions of temperature, flows of heat etc., it becomes possible to create a model of the atmosphere that in many respects resembles our expectations. *The problem with this approach is that the atmospheric model cannot be used for forecasts of future changes.*

This is due to the fact that these highly important conditions at the surface of the ocean are created by the interaction of ocean and atmosphere. A change in the atmosphere for example would lead to a change in the ocean which would in turn give rise to a change in the conditions prevailing at the surface of the ocean. This type of feedback mechanism is of decisive importance for changes in the climate over the long run. Even relatively rapid changes such as the widely discussed fluctuations in the Pacific ocean (El Nino) or the North Atlantic oscillation (NAO) are an indication of the complicated feedback mechanisms between the ocean and the atmosphere.

# The movement of warm currents towards the poles

The earth's climate is only to a lesser extent determined by local conditions. This means that incoming solar radiation at any given latitude is not normally balanced by outgoing solar radiation at the same latitude. Heat is redistributed by complicated circulatory processes in the atmosphere and oceans, creating a more even distribution of temperature throughout the world. Approximately half of the transfer of heat from tropical and sub-tropical areas to the poles takes place in the oceans as warm surface waters tend to move towards higher latitudes. There would therefore appear to be good grounds for taking a closer look at the characteristics of this upper layer of the oceans.

# The upper layer of the oceans

The world's oceans consist largely of cold water between 0 °C and 4°C and a salinity close to the average of 34.7 parts per thousand i.e. one cubic metre of ocean contains 34.7 kg. salt. On top of this huge, almost homogenous body of water, there is a layer of water that has different characteristics from the rest of the ocean. It is in this upper layer of the ocean that the most important things happen. The most decisive factors, the transfer of heat from warmer latitudes and its exchange with the atmosphere all occur here in this upper layer of the ocean.

Due to the higher temperature (max. ca. 28°C), the upper layer of the ocean is lighter. As a result of net evaporation, salinity is also higher (max. 37 parts per thousand). The higher salinity increases density which compensates for part of the temperature effect. The temperature and salinity may be said to compete for control over the density of the surface water. This competition provides the system with very special features that are not easy to manage. We will return to this below.

As a result of the combined effect of temperature and salinity, the density of the surface water is a few parts per thousand lower than the density of the water below. Due to the size of the system, this modest difference in density is sufficient with the help of gravity to govern the dynamics of the movements of the surface of the oceans.

# The dynamics of the oceanic surface layer

The temperature and density of the ocean change continuously in relation to greater depth and higher latitude. This distribution may be given a characteristic depth (H) and density deviation (D).

As a result of its lower density, the surface layer seeks continuously to spread over the entire surface of the ocean. This creates a "thermohaline" circulation whereby the warm surface waters of tropical and sub-tropical areas spread out towards higher latitudes where as a result of cooling, it leaves the surface. This thermohaline circulation provides the oceanic share of the transfer of warmth towards the poles.

This dynamic mechanism for distribution of ocean temperature and density is complicated by the rotation of the earth. As a result of this rotation, movements of the ocean surface mainly take place along the borders of the surface layer. The Gulf Stream follows this type of boundary between the cold waters in the north west and the warm Sargasso Sea.

In spite of major complications, the dynamics of these movements in the ocean surface are relatively well known. For instance, it is known that the transfer of warm water to higher latitudes in the ocean surface layer is proportional to the squares of the variables H and D that characterise the ocean. It follows that if these variables are not estimated with great care, we will obtain highly erroneous estimates of heat transport in the ocean. For example, a ten per cent over-estimate of the values of H and D would overestimate heat transport by approximately fifty per cent.

Unfortunately, estimates of the ocean surface i.e. estimates of the values of H and D are hardly a simple matter. Firstly, the depth of the surface layer, D, increases as water is drawn up from deeper levels by the turbulence of the ocean. Without this mixing of water from different layers of the ocean, the warming of the ocean would be restricted to a relatively thin layer of water that would be quite inadequate as a means of heat transport. Our knowledge of these mixing processes is far from complete.

The other major difficulty concerns the interaction between the surface of the ocean and the atmosphere which determines the temperature and salinity of the ocean surface. Available methods are based on empirical formulae that for example seek to determine the heat flux from the difference in temperature between the surface of the ocean and the atmosphere at a particular level. These formulae are exceedingly primitive and moreover subject to considerable uncertainty.

In conclusion, this brief appraisal of the difficulties indicates clearly that the critical variables, H and D cannot be calculated with sufficient accuracy.

Our argument has so far avoided the serious complications associated with the distribution of salinity. At first sight, it is tempting to assume that the distribution of salt might be of little importance since neither evaporation nor heat transport is affected directly by the salinity of the ocean. This is however a rather hasty conclusion as is shown by the following example:

- Salinity affects the thermohaline circulation through its effect on the density of the ocean surface. In certain areas such as the northern Pacific, the thermohaline circulation is largely eliminated by the effect of salinity on the density field in the ocean.

- In the cold oceans, sea ice will only form when a layer of the ocean close to the surface has a somewhat lower salinity. The existence of this layer allows the temperature of the surface water to fall to freezing point and ice will form despite the lower levels of the ocean having a substantially higher temperature . Here there is a difference between the Arctic and the Nordic seas. In the Arctic basin, as a result of the lower salinity of the surface layer, there is a thick sheet of perennial ice. In the Nordic seas, the water is almost homogeneous right up to the surface which prevents the formation of ice.

114

Let us therefore examine the special difficulties associated with estimates of the distribution of salinity in the oceans.

## Temperature versus salinty - competition with complications

We have already mentioned the competition between salinity and temperature in the surface layer of the ocean. The problem would be relatively simple if the effect of this competition could be limited to the role played by salinity in partially eliminating the effect that temperature has on density. However the great challenge is associated with the lack of "local control" over salinity.

The temperature of the ocean surface is determined by a series of complicated but benign feedback mechanisms. An increase in temperature gives rise to a loss of heat via evaporation and radiation that moderates the initial rise in temperature.

The salinity of the ocean surface, on the other hand, is subject to quite different conditions. The distribution of salinity is created by evaporation and rainfall. An unusually high level of salinity on the ocean surface has a negligible effect on the processes governing evaporation and rainfall. A local surplus of fresh water must then be balanced by a transfer of water to other regions that have a deficit.

This lack of a local stabilising feedback mechanism which characterises the forces governing the salinity field represents a major complication that has far-reaching consequences for the modelling of the sea and its interaction with the atmosphere in so-called climate models.

A particularly troublesome conclusion is that the state of the sea and consequently the climate is not necessarily unique- *the same external conditions may give rise to completely different states*. This was pointed out for the first time by Henry Stommel (1961), a great thinker and pioneer within oceanography with the help of an elegant analogy. We will examine Stommel's conclusion below in our discussion on the Nordic seas.

Another awkward consequence of the dynamic process that controls salinity is that the system may be expected to contain very long time scales. For instance, it may take several thousand years to even out differences in salinity between the great basins of the world's oceans.

## Circulation in the Nordic Seas

The ocean north of Iceland - the Nordic Seas - is an excellent example of an area where we can study the competition between temperature and salinity as the conflicting driving forces underlying circulation.

Ever since the last Ice Age, about 10,000 years ago, the area north of Iceland has been dominated by the present prevailing type of circulation. Relatively warm, salt water forces its way between the Faeroes and Shetland and then up the Norwegian coast towards Spetsbergen. During this passage, the water is cooled to a temperature close to freezing point. This water largely fills the ocean basin north of Iceland before it runs back into the North Atlantic.

A prerequisite for this type of circulation is that the water in the ocean basin is heavier than the water pouring into the system. In connection with the cooling , a certain amount of dilution by fresh water will take place. This will lower the density of the water in the ocean basin and create difficulties for the thermohaline circulation.

Assume that the circulation had over a long period been slowed down by for example abnormal winds. As a result, rainfall would have a longer period in which to dilute the incoming water and thereby lower the salinity of the basin water. Assume that this process reached the point where the water in the ocean basin was no longer heavier than the incoming water from the North Atlantic. The circulation of water would then cease and perhaps not be able to start up again. Hence the possibility may arise that the system may become locked in some type of reverse circulation. Such scenarios are fascinating for physical oceanographers and modellers. Hence it is hardly surprising that numerous scientific articles have been written on this topic during the past decade.

Paleo-oceanographical studies also show that circulation in the Nordic Seas has varied substantially during for example the last Ice Age, particularly toward the end of that period about 10,000 years ago.

It is less fortunate however that these undoubtedly dramatic conditions should have been allowed to act as the basis of nightmare scenarios for climate change in the mass media.

## The Gulf Stream in a warmer climate

The following sentence has crept into the IPCC's latest report (1995):

> "Examples of such non-linear behaviour include rapid circulation changes in the north Atlantic..."

Here it would seem that the IPCC has been unable to resist the temptation to take up the highly speculative ideas about the possibility of global warming creating an "ice age climate" in northern Europe. The idea is that a warmer climate would lead to heavier rainfall and a consequent dilution of the surface waters of the Nordic Seas. This dilution would in turn prevent the Gulf Stream from entering the Nordic Seas and thereby creating a substantially colder and less hospitable climate in northern Europe.

I would not wish to suggest that such a course of events is impossible. The system is so complicated that it would not be unreasonable to state that almost anything may happen. However, as I will argue below, there are stronger grounds for believing that a warmer climate would actually stabilise circulation in the Northern Seas.

The simple narrative below is based on the following strategy: Let us assume that circulation is not drastically changed. Given this assumption, an estimate is made of how the forces that determine the prevailing circulation will be affected by a warmer climate. If this estimate is able to indicate with a certain degree of robustness that these forces are likely to become stronger , it would appear reasonable to assume that circulation will continue to prevail in the warmer climate.

The difference in density between the surface water south of the Nordic Seas and the waters of the basin itself is the principle force generating the inflow of water from the North Atlantic. Estimates of this difference in density must take into account the effects of higher temperatures and changes in salinity.

At present, the temperature of the incoming water is around 8°C and salinity is about 35.3 per mille whereas the temperature of the water of the basin is about freezing point and the salinity is 34.92 per mille. This creates a difference in density of about 0.52 per mille, consisting of 0.82 per mille due to temperature and - 0.30 per mille due to salinity.

Given a warmer climate, let us assume that the temperature rises by one degree to 9°C and 1°C respectively. The heavier rainfall of a warmer climate is likely to raise the difference in salinity but by how much is uncertain.

We expect that a warmer atmosphere will contain more water vapour, approximately 6 per cent more per degree Celsius to be precise. I assume that this will lead to about 6 per cent more rainfall which is almost certainly an overestimate. If the system's function is largely unchanged, this implies that the difference in salinity between the incoming water and the water in the basin also increases by 6 per cent. On the basis of these assumptions, the difference in density will rise from about 0,52 per mille to 0,60 per mille. This difference in density is largely attributable to the rise in temperature, from 0.82 per mille to 0.92 per mille, whereas the negative effect of the difference in salinity changes from - 0.30 per mille to - 0.32 per mille.

These rough estimates would suggest that if nothing unexpected occurs, we can anticipate an increase in the inflow from the Gulf Stream into the Nordic Seas.

Can we reasonably expect that the Gulf Stream will continue to exist or are there other threats? Anders Stigebrandt, professor of oceanography at the University of Gothenburg has presented some interesting ideas in relation to this question (1995). He argues that the predominant cyclonic winds over the Nordic Seas are of decisive importance since they protect the waters of the basin from receiving excessive amounts of melt water from the Arctic etc. As a result of these cyclonic winds, the melt waters are removed from the inner parts of the basin and are also driven out of the Nordic Seas at the southern point of Greenland. It is obviously of the utmost importance to examine how the winds over the Nordic Seas are monitored.

## Some ideas on the climate and its modelling

In the discussion above, I have pointed to a number of difficulties:

- We have inadequate knowledge of the processes that determine the composition of the surface layer of the ocean, the mixing between different layers of the ocean and the exchange between the surface of the ocean and the atmosphere.

- The distribution of the salinity in the ocean is controlled in such a manner that we obtain very long time scales and a lack of uniqueness.

In my view, these problems raise serious difficulties for modelling of the climate in the foreseeable future.

It seems obvious to me that we have to have knowledge of how the climate "works" before we can make forecasts about future changes in the climate. A first insight into the workings of the climate could be gained if we had been successful in producing a climatic model that was able to approach the prevailing climate from a different state. Such a model would have to include processes that provide basic stability to the climate. It is essential to determine these processes before attempting to predict future changes in the climate. The sensitivity of the climate is an inherent part of these processes.

This type of investigation requires the use of extremely complex computer models over very long periods of time. This is partly due to the long time scales required to establish the distribution of salinity in the oceans. Present day computer models do not actually allow us to carry out such analyses.

When devising computer models, one has instead been forced to find other routes that are more fruitful in terms of "results". Instead of developing a model on the basis of an analysis of the present day climate, attention is concentrated on a model that is able to capture the characteristics of the present day climate. This means that one starts off from estimates of present day conditions in the sea and atmosphere. The model is then considered a "success" if it doesn't deviate from the present day climate. The "disturbed climate" of the

118

future is then obtained by allowing this model to operate on the basis of changed assumptions.

This approach is not particularly reassuring. It is my conviction that one should at least have elementary knowledge of both the model and the dynamic characteristics of the actual climate - and above all, the underlying mechanisms that produce the stability of the climate. In other words, prior to making any forecasts about the future, we have to try to understand how the earth's climate has functioned over millions of years without running out of control.

## Final comments

I wish to conclude by putting forward three theses that summarise my own views on the "climate issue".

1. We don't know whether or not it will become warmer as a result of the anthropogenic emissions of greenhouse gases.

2. Nor do we know whether or not global warming will have a damaging or beneficial effect.

3. In practice, we have little or no opportunity to do anything about it.

Points 1 and 2 above may be illustrated with the help of two quotations written by Emeritus Professor Bert Bolin (1996).

1. "The basic idea is that the shortcomings that were present in the models that we used to estimate the natural climate will also affect our estimates of climate change. The difference between these types of estimates should provide a fairly clear picture of the sensitivity of the climate......"

*My comments:*
Unfortunately, it would seem that this "basic idea" has never shown itself to be tenable since the "climatic sensitivity" of the model has not been verified by real climate changes.

2. "A warmer climate will bring about a more intensive evaporation and 4 – 7 per cent more water vapour in the atmosphere for each degree of temperature increase. It is likely that this will lead to drought in some parts of the world and to heavier rainfall in others."

119

*My comments:*

It is perhaps possible to speculate in the following more positive manner: In a warmer climate, the atmosphere will be able to transport more water in over the continents. This should provide a larger part of the continental land masses with life-supporting rainfall.

3. Regarding the possibility of "doing something", there are grounds for concern. Since the climate issue is overwhelming in scale, any half-baked project can appear on the scene to "save the climate". The nuclear industry will almost certainly attempt to use the opportunity to seek generous amounts of largesse from the Ministry of Industrial Affairs. After all what is "a small amount of nuclear waste" in comparison to a climatic catastrophe. The struggle for oil resources can in the future be disguised as a religious crusade to teach the oil producers the correct climatic beliefs. Yes there are grounds for concern. On the other hand, I don't believe for a moment that we will be able to prevent or influence the exploitation of fossil fuels.

## References

Bolin, B. (1996), Ur Naturvetenskapliga forskningsrådets årsbok, s. 14 resp 17.

IPCC (1996), *IPCC Second Assessment - Climate Change 1995*, punkt 2.12, s. 6. IPCC, Genève.

Stigebrandt, A. (1985), On the hydrography and ice conditions in the northern North Atlantic during different phases of a glaciation cycle. *Paleogeograpgy, Paleoclimatology, Paleoecology*, 50: 303-321.

Stommel, H. (1961), Thermohaline convection with two stable regimes of flow. *Tellus*, 13: 224-230.

# Energy &
# Environmental Policy

# 10.

# An economic analysis of climate policy

*Marian Radetzki*

As recently as ten years ago, there was little interest in the human causes of climate change. It was an issue that was primarily discussed by meteorologists with occasional contributions from physicists, chemists and other natural scientists. Subsequently, environmental activists also involved themselves in a debate that had by then become highly inflamed. Allegations that the human emissions of greenhouse gases would lead to dramatic and catastrophic climate changes have attracted the attention of the mass media and politicians during the 1990s. There has been a continuous stream of drastic and costly proposals to eliminate the threat of the greenhouse effect in spite of a distinct lack of clarity surrounding the nature of the threat and the effectiveness of the counter-measures. The conference at Kyoto in 1997 was a clear indication of this situation.

Until recently, the dominant view of the greenhouse effect has been that human interference with the climate is inherently negative and must be stopped regardless of the costs. Economic considerations have been of relatively little importance in this discussion.

However the situation would appear to be undergoing rapid change. A growing number of economists have entered the debate and brought attention to the economic issues involved in the debate on the greenhouse effect. As has been the case with the natural scientists, the economists have divided themselves into different groups. A minority of economists consider that future changes in the global climate constitute a disastrous threat and that immediate counter-measures are economically justified. Another group argues that climate change, if it actually takes place, will have limited economic consequences while the costs of reducing emissions of greenhouse gases will be on a major scale. Hence in the view of this group, it would not make economic sense now to invest heavily in measures to reduce greenhouse gas emissions.

The UN's International Panel on Climate Change has explored many of the arguments on the climate issue that are currently the subject of debate among economists. Its analysis of the issues has been summarised in a recently published volume (IPCC, 1996). However other important contributions have been published both before and after the appearance of the IPCC volume. IPCC (1996) consists of a collection of contributions from a number of authors. Hence a comparison between chapters or indeed even within the same chapter indicates a range of views and opinions that are not always consistent with each other.

I have not myself made a contribution to the scientific literature on the climate issue. However I am an active student of the literature. The aim of this chapter is to use my insights to identify the economic aspects of the climate issue, to examine the analytical arguments employed by economists and to present some of the findings from the economic analyses. In these parts, my discussion is heavily dependent on the IPCC volume. In the final section, I draw some conclusions regarding the central question of the type of climate policy that ought to be adopted on economic grounds, given our present state of knowledge.

## Uncertainties

The impact of human activity on global climate is an issue that is surrounded by considerable uncertainty. It is evident from the contributions made by my colleagues in this anthology that science has by no means a tight grip on the central relationships and it is therefore difficult to make predictions. Two major sources of uncertainty are piled on top of each other: (a) the extent to which greenhouse gases are actually end up in the atmosphere and the length of time that they remain there and (b) in spite of the availability of increasingly sophisticated climate models, substantial uncertainty surrounds the effect on the climate of growing concentrations of greenhouse gases in the atmosphere.

The economic effects could be said to be the final link in the chain. The economic analysis which is itself subject to substantial uncertainty must be based on the scientific uncertainties disputed by meteorologists, physicists and chemists. For instance there are differences of view on the actual extent of future greenhouse gas emissions in the absence of any special measures to reduce such emissions. The economic analysis of the utility to be gained from a stabilisation of the global climate is subject to a high degree of uncertainty regarding both the nature and extent of the utility gains. There are also

123

substantial variations in the assessment of the costs of reducing emissions of greenhouse gases over time, a principal measure for stabilising the climate. Both costs and benefits occur in the future although there is a lack of agreement as to exactly when they will occur. For a proper comparison, the flows of costs and benefits over time have to be discounted to provide an estimate of their present value. Here there is also disagreement regarding the discount rate to be employed. The results of the economic analysis are however highly sensitive to the choice of discount rate.

The extremely wide range of possible economic repercussions that are a consequence of these uncertainties allows substantial scope for differing views regarding an appropriate climate policy. Those who recommend measures aimed at a far-reaching stabilisation of the climate are able to draw support from the drastic economic effects that would arise from an absence of measures to deal with the threat to the global climate. The contrary view that no measures need to be taken is able analogously to draw support from the opposite scenario where the anticipated climate changes are not seen to be particularly dangerous and indeed may even be beneficial at the same time as the costs of reducing greenhouse gas emissions are high. The agnostics, with whom I would wish to be associated, argue in favour of further analyses that will hopefully reduce the range of outcomes. Until then, the agnostics are only prepared to support a modest, relatively inexpensive range of climate policy measures.

## Two points of departure

### Cost-benefit analysis and discounting

An economic evaluation of climate policy projects requires a benchmark. The absence of measures to reduce emissions provides us with such a benchmark, namely what the IPCC terms the "business as usual" (BAU) scenario[1].

The purpose of climate policy is to create *benefits* by preventing damage to the climate i.e. undesirable climate changes that might be expected to arise in the BAU scenario. Hence the value of the benefits created would be equivalent to the costs of the damage to the climate that are avoided as a result of the climate policy measures. Implementation of climate policy, however, involves a variety of costs. A realistic analysis has obviously to take into account both benefits and costs.

---

[1] BAU is not a concept that offers a straightforward definition. The IPCC has worked with a number of BAU scenarios for several years. In the discussion below, I refer to the scenario that the IPCC has termed IS 92a. However it should be noted that this scenario appears to have been recently abandoned by the IPCC.

Cost-benefit analysis constitutes an appropriate economic tool for this evaluation exercise. An investment project that gives rise to a stream of future costs and benefits will only be justifiable on economic grounds if the present value of the total benefits that flow from the project exceeds the present value of its costs. Measures to stabilise the climate may be evaluated in this way. In order to quantify the benefits of climate stabilisation and the costs of the measures involved, economists would first wish to select a scientifically sound climate model and integrate it with a model describing the operations of the global economy. Taking into account as far as possible all relevant factors, climate policy measures would only be justified when the present value of the benefits of the project exceeds the present value of its costs. This approach is simple, clear and incontrovertible in principle. However, as we shall see below, in practice, this approach leaves a lot to be desired in terms of simplicity and clarity.

My view of cost benefit analysis as an essential tool for the evaluation of climate policy projects is not particularly controversial although it is not shared by all economists working in this field. To quote the IPCC:

"Despite the current limitations of these various techniques, modern CBA (broadly defined) remains the best framework for identifying the essential questions that policymakers must face when dealing with climate change. The CBA approach forces decision makers to compare the consequences of alternative actions, including that of no action, on a quantative basis... the most important benefit of applying CBA is... the process itself (which establishes a framework for gathering information and forces an approach to decision making that is based on rigorous and quantitative reasoning)."

It is also true that the IPCC presents an alternative to cost- benefit analysis, e.g. the *sustainability* method recommended by certain economists. However it is not possible to use this method in order to carry out a rational economic analysis since its proponents consider it to be axiomatic that climate damage will be overwhelming and that stabilisation of the climate will be necessary irrespective of the costs involved (IPCC, 1996, pp 184-185).

When a project is long-run and especially where the benefits arise at a much later date than the costs, the economic outcome will be heavily dependent on the rate of discount that is used to estimate its present value. Normally the rate of discount ought to reflect the long run market valuation of the invested capital. If the discounted present value of the project's benefits is less than its discounted costs, the project will not be able to cover its capital costs. This indicates that the project is unprofitable and that it would be more advantageous to invest in other projects where the expected return is higher. The majority of economists working in the field of climate policy apply these principles.

There are however divergent views on this issue. Some economists contend that the overwhelming importance and extremely long run nature of climate problems would argue in favour of discounting the benefits of climate stabilisation at a very low, possibly zero, rate of interest. This would create problems. The shortage of investment resources always gives rise to difficult choices between numerous competing priorities. Other economists may also legitimately argue that the climate problem is not unique in terms of importance or long term effects and that preferential treatment is equally warranted for many other pressing projects.

The rate of discount provides a discipline and a choice criterion that can be used when it is not possible to satisfy all of the competing demands for investment resources. Serious mistakes can be made when this discipline is not applied. The leaders of the Soviet Union failed to take account of the rate of return to capital when they planned their investment decisions. They were convinced that they were best equipped to decide where the available capital resources would produce the greatest social returns. It is obvious now in retrospect that these decisions were not always the correct ones. The burden of evidence rests heavily on those who claim that the climate problems are so special that they don't need to be subjected to the discipline of the capital market.

The choice of the discount rate would not present any difficulties if the special character of the climate problem comprised a likely potential catastrophe that would completely destroy mankind. The damage resulting from such a climate catastrophe would give rise to infinitely large costs for humanity. The present value of these costs would remain infinitely large irrespective of the timing of the catastrophe or the choice of discount rate.

An example of the use of cost-benefit analysis in relation to the climate issue is illustrated in Figure 10.1 which shows the estimated global benefits and costs over a 280 year period of an ambitious programme of climate stabilisation (Cline, 1992). The annual benefit of this programme is expected to increase continuously from a level of just over 1 per cent of global GDP around the year 2050 to almost 5 per cent in the year 2270. These figures reflect the gradual and increasingly substantial deterioration in the unregulated global climate. In the early stages, the annual costs of implementing the programme amount to between 3 and 4 per cent of global GDP. However this figure tends to fall as the world economy adjusts to the climate policy constraints and GDP rises. It is estimated that by approximately 2160, the annual flow of benefits will exceed annual costs.

A visual inspection of the climate policy measures in this example is revealing. It is quite evident that the sum of the benefits accruing to the project will be lower than the costs of the project. This conclusion holds even when the rate of interest is set at zero. The present value of the project will be strongly negative even at low discount rates since the benefits of the project accrue

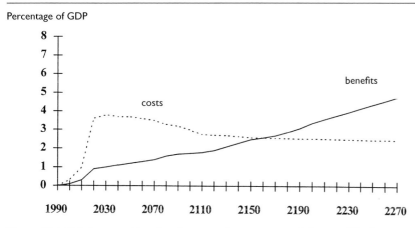

Percentage of GDP

**Figure 10.1** Global costs and benefits of an aggressive programme of climate stabilisation.

relatively late whereas costs are substantial at an early stage of the project.

This type of aggressive programme cannot be justified in economic terms. If the programme's costs and benefits are an accurate reflection of human values, it is obviously less troublesome to adjust to the deterioration in climate than to pay to avoid it.

There are actually only two ways in which the above programme could be justified on economic grounds. The first would be to extend the time horizon while at the same time applying a very low discount rate. If the annual cost continued to fall beyond the year 2270 while benefits grew along the previous trend, one would reach a point where the present value of total benefits exceeds the present value of total costs. The second way would be to replace the baseline case for estimating the benefits of the climate policy programme with another scenario where the benefits from climate stabilisation measures, i.e. the damage to the climate as a result of a BAU policy, are higher than shown in the figure.

Although Figure 10.1 is only an example, it is instructive nevertheless to see how cost-benefit analysis is applied and how this analytical method can provide a basis for rational economic decisions. The underlying assumptions may certainly be varied ad infinitum and consideration can also be taken of factors that are inherently difficult to evaluate. However in the absence of cost-benefit analysis, it would simply not be possible to deal with the climate issue in an economically rational manner.

127

## A doubling of the atmospheric concentration of carbon dioxide

Climate policy discussions often have a doubling of the atmospheric concentration of carbon dioxide as a benchmark variable[2]. The dominant view during the late 1990s is that such an increase in carbon dioxide would raise global temperatures by 2.5°C within a broad confidence interval. Many climate policy projects set their targets for reducing emissions to a level that would prevent the atmospheric concentration of carbon dioxide and the increase in temperature from exceeding these values. The aim of Table 10.1 is to provide a framework for this benchmark and objective and to place the Kyoto conference exercises regarding emission reductions in a broader perspective.

The pre-industrial atmospheric concentration of carbon dioxide amounted to 280 parts per million (ppm). As a result of human emissions, it had increased to 360 ppm by the 1990s. Annual emissions of carbon dioxide attributable to human activity have increased steadily and have now reached a level of 7.3 billion tons of carbon, of which 1.6 billion tons is the result of deforestation while the remainder is due to the burning of fossil fuels. Although the impact of these emissions on global temperature is unclear, a number of guesses put the increase at about 0.5°C. In the absence of climate policy measures, the atmospheric concentration of carbon dioxide is expected to double by about 2050, by which time the human emissions of greenhouse gases may have risen to 16 billion tons of carbon.

A conventional view is that emissions should not be permitted to exceed an annual average of 7 billion tons of carbon between 2000 and 2100 and that they should fall to under 5 billion tons per annum during the following century in order to prevent the atmospheric concentration of carbon dioxide from exceeding a level of 560 ppm. On this basis, already in 25–30 years, emissions would have to be 40-50 per cent below the BAU level. This would require far greater emission controls than would be achieved by a linear projection of the Kyoto decision.

---

[2] Other greenhouse gases account for about a third of the total greenhouse effect caused by human activities. For the sake of simplicity, only $CO_2$ is dealt with explicitly here. However the results of the analyses presented below of the benefits to be derived from climate stabilisation and the costs of reducing greenhouse gas emissions include the other greenhouse gases in terms of their $CO_2$ equivalents.

**Table 10.1** Atmospheric concentration of carbon dioxide, emissions and effect on temperature

|  | Concentration Ppm | Emissions billion t C | Increase in temp. °C |
|---|---|---|---|
| "Pre-industrial" | 280 | – | – |
| 1990s | 360 | 7.3 | 0.6? |
| 2050 (BAU) | 560 | 16 | 2.5 |
| | | | |
| Goal | | | |
| 2000 –2100 | <560 | 7(average) | <2.5 |
| | | | |
| Sub-goal | | | |
| 2020 –2030 | | 7 (40–50% below BAU) | |

Source: Cline (1992): SOU (1995)

Billion tons Carbon

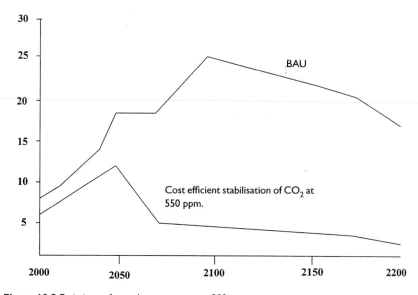

**Figure 10.2** Emissions of greenhouse gases over 200 years

Figure 10.2 (IPCC, 1996, p.338) compares the expected increase in emissions until 2200, assuming BAU, with another project for emission reductions that seeks to keep the atmospheric concentration of carbon dioxide below double

129

its pre-industrial level[3]. According to this project, emissions continue to increase for the next 50 years but are subsequently drastically lowered to a level below 5 billion tons in 2075. It is possible that this type of project may be more cost-efficient and it is presumably the basis on which the Kyoto agreement was reached. Difficult and costly decisions are simply postponed. If the Kyoto agreement is put into practice, the industrial countries that presently account for about half of global emissions of greenhouse gases (World Bank, 1997) would marginally reduce their absolute emission levels during the next decade while other countries would continue to rapidly increase their emissions. It would not be until around the middle of the next century that drastic and extremely costly measures would be undertaken. These reductions in emissions would also place a great burden on the third world, which would have to cooperate fully if there is to be any possibility of reaching the target level.

## The benefits of climate stabilisation

The benefits of stabilisation are equivalent, as we have previously stated, to the damage to the climate in the BAU scenario that we avoid by undertaking measures to reduce emissions.

Various assessments have been made of the damage that would be likely to occur if $CO_2$ concentrations reached 560 ppm and global average temperatures rose by 2.5°C. Most of these estimates relate to individual nations. However there are two estimates that were extended to cover the entire globe (IPCC, 1996, p. 203-205). The global estimates by Fankhauser (1995) and Tol (1995), that are quoted by the IPCC, indicate an annual cost of between 270 and 320 billion dollars or 1.4 – 1.9 per cent of world GDP in the late 1980s. A regional distribution shows, unsurprisingly, that the rich economies are less sensitive to climate change but that poorer countries will be more affected. Especially low levels of climate damage will be experienced by the former Soviet Union where the climate is initially quite cold.

The estimated damage caused by climate change would appear to be based on the GDP and economic structures that existed around 1990. In actual fact, this climate damage will occur during the latter half of the twenty first century since it will not be until then that the concentration of carbon dioxide in the atmosphere will have doubled in the absence of appropriate counter-measures. Since a gradual global warming should provide warning signals of approach-

---

[3] Note that the BAU scenario IS 92a only extends to the year 2100. The reduction in emissions during subsequent years is based on an extrapolation carried out by the IPCC under the assumption that fossil fuels will be gradually exhausted.

ing problems and offer plenty of time for adjustments to be made, all while global GDP can be expected to grow substantially during the next fifty years, it could be argued that future climate damage will actually comprise a significantly lesser proportion of GDP than quoted above. Tol himself estimated that an increase in average global temperatures of 2.5°C per cent occurring around 2050 will give rise to environmental damage equivalent to 0.55 per cent of world GDP (IPCC 1996, p. 209) which is only one third of the level indicated above.

Climate damage may be expected to be negligible in the next few decades assuming that emissions of greenhouse gases increase along the lines outlined in the BAU scenario and if global temperatures rise in proportion to the increased atmospheric concentration of carbon dioxide. If we are able to accept the above estimate, the damage attributable to climate change should be equivalent to about 0.5 per cent of GDP in the middle of the twenty first century. However global warming will continue. In an extreme case, it has been argued that average global temperatures could increase by 10°C over the next three hundred years. Even assuming a substantial degree of adjustment to such climate change, climate damage would be likely to increase exponentially. Each degree increase in temperature would give rise to a growing absolute increase in the overall level of climate damage. However as a result of the expected exponential increase in world economic growth, it is even then uncertain whether climate damage will increase relative to global GDP.

The possibility cannot be excluded that greenhouse gas emissions will lead to a climate catastrophe, producing widespread damage on a scale not previously envisaged. The IPCC (1996, pp 208-209) mentions three possible catastrophe scenarios:

• A galloping greenhouse effect where the original increases in global temperatures produce secondary effects that reinforce global warming e.g. the release of large amounts of methane gas from a thawing tundra.

• A large scale smelting of the Antarctic that raises the ocean level by 5-6 metres.

• A blockage of the Gulf Stream bringing about dramatic reductions in temperatures in Northern Europe.

The IPCC does not give much space to a consideration of these catastrophes. No attempt is made to estimate the probability of any of these catastrophes or to assess their economic consequences. Indeed the IPCC does not even indicate a clear relationship between emissions of greenhouse gases and the likelihood of such catastrophes. Hence it cannot be stated with any certainty that

131

the probability of such catastrophes will increase as a result of climate change. Similarly it is by no means certain that the envisaged measures to stabilise the climate will reduce the risk that such catastrophes will occur.

In spite of the prevailing uncertainty, a number of attempts have been made to identify the contours of an optimal climate policy over long periods of time. The approach has been to integrate the value of the anticipated total climate damage at various levels of emissions over time and to estimate the present value of this integral. The estimated value of the damage from potential future climate events has been adjusted to take account of the likelihood of their occurrence. The outcome of the estimates of climate damage is then compared to the emissions of carbon dioxide over time in order to estimate the marginal cost to society of emitting an additional ton of carbon into the atmosphere. If the polluter is then taxed at an amount corresponding to the marginal social cost of his carbon emission, he may be said to have recompensed society for the environmental damage that he caused. The carbon tax provides in a simplified form a measure of the marginal utility of climate policy.

Five independent estimates of a socially optimal climate policy are presented by the IPCC (IPCC, 1996, p.215). The results are summarised in the left-hand column of Table 10.2. As can be seen, the benefits of reduced emissions increase over time. This is due to the fact that climate damage increases more than proportionately for each degree increase in temperature. The benefits of preventing climate change, measured in tons of carbon that are not emitted will consequently be greater in 2095 when the temperature has already risen, than a century earlier.

With one exception (see below), the studies show relatively low figures for fossil fuel taxation. Hence the reduction in emissions will be correspondingly low. In the optimal climate policy, levels of greenhouse gas emissions are only 5-10 per cent below the level that would have prevailed in the BAU case at the end of the twentieth century and 10–30 per cent below the BAU benchmark one hundred years later (Fankhauser, 1994; Nordhaus, 1994; Tol 1994). It is apparently easier to adjust to climate change than to meet the costs of measures to stabilise the climate.

The total annual benefits of the optimal climate policy implicit in these studies can be estimated with the help of these carbon taxes. This is shown in columns 2 and 3 of Table 10.2. For two reasons, however, these figures tend to overestimate the benefits. *Firstly*, I have assumed for the sake of simplicity that the marginal benefit of emission reductions is equal to the average benefit. This overestimates the gains to society of a climate policy programme. If greenhouse gas emissions have already been substantially reduced, the marginal utility of an additional reduction ought to be markedly lower than the initial gains from an emissions reduction programme. *Secondly* I have assumed throughout that global GDP will increase by 2.5 per cent per annum. In terms

132

of the twentieth century, this is an extremely low figure. If GDP growth is more rapid, the benefits from climate stabilisation will be less, expressed as a percentage of GDP.

**Table 10.2** The benefits of climate stabilisation

| Year | $ per ton reduction in carbon emissions | Total $ bill. | Percent of GDP |
|------|------|------|------|
| 1995 | 5–20 | 37–146 | 0.1–0.5 |
| 2025 | 10–28 | 120–336 | 0.2–0.6 |
| 2095 | 21–90 | 500–2,200 | 0.2–0.7 |

It should be noted that the figures presented by Cline (1992) diverge from the above estimates. Although one of his two estimates for the dollar value of the reduction in carbon emissions falls within the range shown in the above table, the other is far in excess of this range of values, namely $ 124 per ton for 1995, $ 221 per ton for 2025 and $243 per ton for 2095. Only with carbon taxes at these high levels could one realistically expect a reduction in emissions that would be sufficient to arrest the increase in the atmospheric concentration of carbon dioxide at a level of 560 ppm.

The basis of this substantial deviation in Cline's second series of figures is not a fundamentally different view of climate change but the artificially low rate of discount that he uses in his present value estimates. On the basis of these extreme values, the benefits of Cline's dramatic climate policy are equivalent to 3.2 per cent of global GDP in 1995, 4.5 per cent in 2025 and 1.8 per cent in 2095.

## The costs of climate stabilisation

The costs of climate stabilisation consist mainly of the added expenditure for replacing cheap fossil fuels with more expensive non-carbon based alternatives. However costs will also arise from the forced reduction of total energy consumption. The introduction of carbon taxes will give rise to a major and long-lasting cost of adjustment. The painful and drawn-out experience of the 1970s when the global economy tried to adapt to higher oil prices is a reminder of the magnitude of the problems that will have to be overcome in conjunction with a serious effort to stabilise climate. A cost-benefit analysis can then be used to compare these costs with the benefits of an emission reduction programme, identified above.

133

The results from various attempts to estimate the global annual costs of climate policy programmes that seek to reduce greenhouse gas emissions are presented in Figure 10.3 Richels and Sturm, 1996. IPCC, 1996 presents a series of similar results from some of the same studies). Each estimate is identified in the figure by the particular year in question, the level of ambition in relation to emission reduction, and the cost expressed as a percentage of GDP. It is hardly surprising that the costs tend to increase as the level of ambition rises.

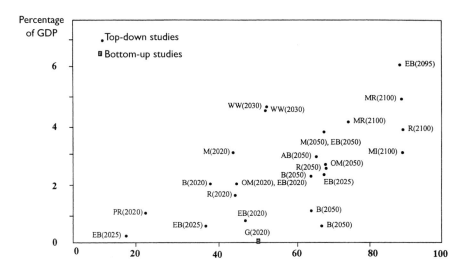

**Figure 10.3** Global costs for the reduction of $CO_2$. The results of 24 studies. Percentage reduction in emissions in relation to BAU at a given point of time.

The results diverge markedly even among studies that have similar levels of ambition. For example, in six of the studies, the costs of the reductions in emissions of between 60 and 70 per cent in relation to the BAU level, vary in the year 2050, from 0.5 per cent to 4 per cent of GDP. Encouraged by the low figures in the range, the supporters of a radical climate policy have argued, especially during the debates that preceded the conference in Kyoto, that substantial reductions in greenhouse gas emissions need not be particularly expensive. Indeed some supporters of this line of argument have suggested that climate policy programmes are not just self-financing but actually provide economic returns in addition to those that result from an improved climate.

The basis for such an argument has been examined in a newly published report from the World Resources Institute (WRI) and is presented here in summary form (Repetto and Austin, 1997). The WRI study contends that the major part of the difference in costs of climate policy measures is attributable to the alternative assumptions that are used in relation to a limited number of factors of critical importance to the level of costs rather than to any difference in the structure of the analytical models. On the basis of no less than 16 different model formulations, the WRI researchers analyse the variation in costs to the US of a far-reaching climate policy programme in the year 2020, dependent on the assumptions used. The conclusions would also appear to be applicable in a modified form at the global level. Table 10.3 presents the results as a percentage of USA's GDP.

Adopting pessimistic assumptions for all critical factors, the costs of reducing emissions will exceed 7 per cent of GDP. As we choose optimistic assumptions for an increasing number of factors, the costs of climate policy will fall. Ultimately, the programme of emission reductions will yield an addition to GDP amounting to 4.5 per cent. In several areas, this growing optimism will require careful scrutiny.

**Table 10.3** Effect on GDP in the USA in 2020 of a 60 per cent reduction in greenhouse gas emissions, compared to a BAU scenario, under increasingly "optimistic assumptions."

| | Outcome for specific factor | Cumulative outcome |
|---|---|---|
| Pessimistic assumptions for all critical factors | | −7.2 |
| 1. Trade with emission rights, national and international | +1.4 | −5.8 |
| 2. Alternatives to fossil fuels in unlimited quantities at constant costs (backstop) | +1.8 | −4.0 |
| 3. Capacity to adjust at macro- and micro-economic level is much larger than historically | +3.6 | −0.4 |
| 4. Fossil fuel taxation will finance the elimination of growth-restricting taxation | +2.6 | +2.2 |
| 5. Positive health effects of lower levels of atmospheric pollution following reductions in the use of fossil fuels | +2.3 | +4.5 |

The optimists assume firstly that climate policy will bring about a greater flexibility in the behaviour of both households and the economy at large compared to previous decades. This greater adaptability will lower the costs of adjustment to the climate policy measures. Even more important however is the fact that a greater measure of flexibility will raise the overall level of efficiency in the economy and allow economic growth to increase at a more rapid rate than it would otherwise have done.

A second assumption made by the optimists is that the high level of taxation of fossil fuels will replace a number of existing taxes that are inefficient and that curtail economic growth. This reform of the tax system will release productive forces and stimulate the rate of economic growth.

Thirdly, it is assumed that the replacement of fossil fuels by other forms of energy will lead to a dramatic reduction of atmospheric pollution, which will substantially contribute to better health.

These three assumptions will have a combined positive effect on GDP amounting to 8.5 per cent. From having been a major burden to the economy, climate policy measures are transformed magically into a measure that is beneficial to the economy, even before we take account of the advantages of a more stable climate. Climate policy is seen to stimulate economic growth and GDP will thereby be several percentage points higher than in the absence of climate intervention.

A closer look however shows that this conclusion is not borne out by reality and has no basis in climate policy. There is every reason to take measures to increase the flexibility of the economy but there are no grounds to believe that this flexibility will be greater as a result of climate policy. Similarly there are good reasons for a rationalisation of the tax system but little evidence to show that this rationalisation would be made easier by new taxes on the use of fossil fuels. The positive effects identified by the enthusiasts are extremely desirable but have little to do with climate policy.

If climate policy is to improve the quality of the atmosphere, oil, coal and gas will have to be replaced by energy forms that produce lower emissions of sulphur, nitrogen oxides, ash and other types of substances that can damage health. Nuclear energy meets these requirements. Solar energy is also a pure form of energy but both technology and the economy would tend to argue against its application on a massive scale during the next twenty years. Bio-energy would appear to have a greater opportunity to expand on a large scale. Atmospheric pollution as a result of the use of bio-energy is however considerably larger per unit of energy consumed compared to the fossil fuel that it replaces. In this perspective, the favourable health effects of climate policy would appear to be illusory.

There are consequently no grounds for the optimist's view that climate policy does not impose any costs. If the objective is to put a ceiling on the atmospheric concentration of carbon dioxide emissions at 560 ppm, twice the " pre-industrial" level, global emissions must be reduced to about half the level that they would have been in the BAU scenario as early as 2050, and subsequently by much more. This type of policy will, conservatively, give rise to a permanent annual cost of at least 2 per cent of global GDP and possibly sub-stantially more. Here we are involved with very large sums. In 1995, world GDP amounted to $28,000 billion. By 2050, at an annual rate of growth of 2.5 per cent, world GDP would be around $100, 000 billion. In 1995, 2 per cent of world GDP would amount to $560 billion and $2 000 billion in 2050. These figures indicate that huge resources are going to be required to reach the climate policy objectives. Such a massive resource commitment cannot be undertaken without a very careful consideration of what is involved.

## Synthesis and conclusions

My discussion of the costs and benefits of climate policy are largely based on the IPCC's own analysis and research findings. The picture that has emerged offers little support for an ambitious policy that seeks to stabilise the climate. The economic benefits would appear to be limited while the costs of the sta-bilisation measures are substantial. Hence cost-benefit analysis does not justify this type of policy.

Relatively small reduction in emissions are achieved by the taxation of car-bon emissions at a level equivalent to the marginal cost to society of such emis-sions. Benefits to society are maximised at a substantially lower level of ambition than that which seeks to put a ceiling on the atmospheric concentra-tion of carbon dioxide at twice the pre-industrial level. The present analysis suggests that it is economically more advantageous to adjust to the approach-ing climate change than to take measures to ensure that it does not take place.

A clear threat of an overwhelming climate catastrophe could act as motive for far-reaching, costly measures to stabilise the climate. Our present state of knowledge does not permit us to identify either the nature or probability of a climate catastrophe. Moreover the relationship between BAU and a climate policy with various contents on the one hand, and a future climate catastrophe on the other, is not clear. Hence it is not possible to identify a climate policy that will be able to reduce the risk of a climate catastrophe.

The climate issue is of enormous importance for mankind. However the far-reaching and costly climate programme discussed above should be delayed until we have a better grasp of the relationships involved. Consequently, cli-mate policy measures should be limited to the less ambitious activities involv-ing lower costs outlined by the IPCC in its background analyses. Continuing ambitious efforts to improve our knowledge of the scientific and economic relationships between the emissions of greenhouse gases, climate changes and consequences for society should be given priority.

Those who argue in favour of immediate, far-reaching and costly climate policy measures are almost always to be found in the rich, industrial world. They often support their arguments on ethical grounds. For instance, they argue that cost-benefit analysis underestimates non-material values, fails to take due consideration of the third world which is only a junior partner in the global economy and discounts costs and benefits to obtain their present value, thereby ignoring future generations.

The following table presents a stylised view of the consequences of climate measures between countries and generations. It is based on the argument, realistic in my view, that the rich countries will have to bear most of the costs of climate policy while the largest benefits will accrue to future generations in the third world. An eagerness to take on the major share of the costs of climate stabilisation may in this schematic view be seen as economic aid from the present generation in rich countries to future generations in less developed countries. (Schelling 1995). This view seems, to say the least, rather paternalistic. The present generation of the less developed world should at least be asked to express its opinion on how the potential funds should be used, particularly in view of the high degree of uncertainty surrounding the benefits of climate policy. It is unlikely that the present generation in the third world would choose climate stabilisation rather than measures to put a stop to a number of pressing contemporary difficulties.

|  | Rich Countries | | Poor Countries | |
| --- | --- | --- | --- | --- |
|  | Benefits | Costs | Benefits | Costs |
| Present generation | None | Large | None | Small |
| Future generation | Small | Large | Large | Small |

The number of saved lives is the most basic, albeit highly incomplete, method to evaluate human benefits. This statement would appear to be difficult to refute on ethical grounds. I ignore for the moment all other kinds of utility and study the outcome in terms of survival when resources are devoted to climate policy or alternatively to the undoubtedly pressing objective to save human beings from a painful, sudden death.

The ambitious climate programme that seeks to stabilise the atmospheric concentration of carbon dioxide at twice its "pre-industrial" level is estimated to require a resource commitment of upwards of 2 per cent of global GDP or at least $500 billion per annum. It is uncertain whether and when and to what extent, this programme will actually save lives. The effects in this respect are likely to be modest and occur well into the future.

Assume that the rich world decides instead to devote the same annual sum to a programme to prevent painful, sudden death. Tens of millions of people,

138

largely in the third world, die annually from malaria and tuberculosis and many more, mostly children die from dysentery caused by infected water in combination with malnutrition. There is no doubt that $ 5000 – $10 000 would be more than sufficient to prevent each such premature death. Hence an annual expenditure of $ 500 billion dollars would immediately save 50 –100 million lives per year.

The advocates of an immediate, far-reaching and costly climate policy would not generally accept the economic arguments against climate interventions. But would they be able, on ethical grounds, to recommend climate stabilisation in favour of a policy of directly saving lives?

# References

Cline, W. (1992), *The Economics of Global Warming*, Institute for International Economics, Washington DC.

Fankhauser, S. (1994), The Social Costs Of Greenhouse Gas Emissions: An Expected Value Approach, *Energy Journal*, Vol 15, No 2.

Fankhauser, S. (1995), *Valuing Climate Change: The Economics of the Greenhouse*, Earthscan, London.

IPCC (1996), *Climate Change 1995 - The Economic and Social Dimensions of Climate Change*, Cambridge University Press.

Nordhaus, W. D. (1994), *Managing the Global Commons, The Economics of Climate Change*, MIT Press, Cambridge Mass.

Repetto R., Austin, D. (1997), *The Costs of Climate Protection: A Guide for the Perplexed*, World Resources Institute, Washington DC.

Richels, R.,Sturm, P. (1996), The Costs of $CO_2$ Emission Reductions, *Energy Policy*, Vol 24, No 10-11.

Schelling, T. (1995), Intergenerational Discounting, *Energy Policy*, Vol 23, No 4-5.

SOU (1995), *Jordens klimat förändras*, Klimatdelegationen SOU 1995:96, Stockholm.

Tol, R. (1994), The Damage Costs of Climate Change: A Note on Tangibles and Intangibles Applied to DICE, *Energy Policy*, Vol 22, No 5.

Tol, R. (1995), The Damage Costs of Climate Change: Towards More Comprehensive Calculations, *Environmental and Resource Economics*, Vol 5, p 353-374.

World Bank (1997), *World Development Report 1997*, Washington DC.

# 11.

# Swedish Climate Policy

*Karl-Axel Edin*

If the level of taxation on carbon dioxide emissions can be seen as a measure of climate policy ambitions, Sweden would be ranked as a world leader. Sweden has actually the highest carbon dioxide tax in the world. Indeed only Sweden and more recently its neighbours have taken any concrete steps to reduce their carbon dioxide emissions. Sweden raises approximately S.kr. 10 billion through its taxation of carbon dioxide emissions. What are the benefits of this tax? Is Sweden, a country that is responsible for only a few parts per thousand of global emissions able to influence the earth's climate?

## The climate issue in Swedish politics

Global climate first became a serious political question in Sweden in 1988. It was then that ruling Social Democratic Party decided to decommission two nuclear reactors. The Conservative Party opposed this, partly on the grounds that decommissioning nuclear reactors would lead to the increased use of coal and oil and to increased emissions of carbon dioxide. Subsequently the Swedish parliament approved a conservative proposal to limit carbon dioxide emissions. This decision to set targets for carbon dioxide emissions in Sweden received the support of the Green Party since such a reduction in carbon dioxide emissions would lead the way to renewable energy sources in Sweden. The Swedish Centre party also supported this proposal.

The introduction of energy taxation in the early 1980s was the next step in the growing importance of climatic issues in Swedish politics. The principal motive was the need to increase government revenue in order to reduce a large budget deficit. This was the principal reason for the major increase in energy taxation that was made in 1991. It was at this point that a separate carbon dioxide tax was introduced as a complement to existing energy taxation.

The introduction of the carbon dioxide tax had actually little to do with the climate issue. The tax was designed to reduce the government's deficit. However the legislative proposal to raise the tax on carbon dioxide emissions contained a reference to the parliamentary decision in 1988 to reduce such emissions.

Hence there are several driving forces underlying climate policy and the taxation of carbon dioxide emissions. Those who were opposed to the phasing out of nuclear power have been able to use climate policy as an argument against decommissioning nuclear power stations. The Ministry of Finance has also been able to use the climate issue to raise taxes. The Green Party and environmental pressure groups within other political parties also saw an opportunity to support the cause of renewable sources of energy by means of higher taxes on oil and coal. Furthermore the Centre Party was attracted by the new market opportunities available to forest owners to sell timber and peat.

Hence different political groups have been governed by different motives regarding the role of public policy in relation to climate issues. An evaluation of Swedish climate policy may therefore be carried out from the standpoint of several different objectives. In this context, the stated official policy goal of reducing emissions of carbon dioxide appears as just one of these objectives. The emphasis here however will be on this official policy goal i.e. reducing emissions of carbon dioxide rather than dealing with other these other aspects.

## Taxation of carbon dioxide emissions

The taxation of carbon dioxide emissions is the single most important measure of climate policy in Sweden. It raises the greatest amount of revenue and has the largest impact on carbon dioxide emissions.

The first question that could be raised is to try to explain why the carbon dioxide tax has been set at the current level of 37 öre per kg. carbon. The principal requirement for an efficient environmental tax is that it should reflect the damage brought about by the emissions. This is not the case with the taxation of carbon dioxide. Here the tax on emissions cannot reflect the damage to the environment, if any, that is caused by these emissions since we have no knowledge of the extent of this damage. It is exactly this question that is occupying researchers all over the world at the moment. So far they have not been able to reach an answer. The level of the carbon dioxide tax has been determined by the need to raise taxes.

# The effects of carbon dioxide emissions

Emissions of carbon dioxide in Sweden are approximately 60 million tons per annum. This is equivalent to 0.3 per cent of global emissions. It is estimated that annual $CO_2$ emissions in Sweden are 10 million tons lower than they would otherwise have been in the absence of energy and carbon dioxide taxation.

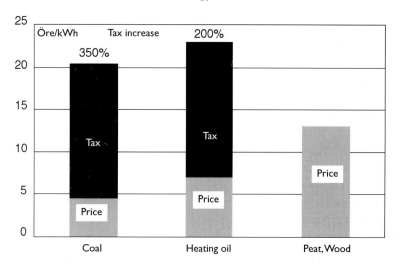

**Figure 11.1** Energy and carbon dioxide taxes on oil and coal are very high in Sweden

This reduction has been achieved at great cost. The question that has to be raised here is whether the country has received the greatest possible return on the money that it has invested in the reduction of emissions.

When evaluating Swedish climate policy, it should be borne in mind that Sweden is by and large the only country that has taken steps to reduce carbon dioxide emissions. As a result the effect of Swedish policy is reduced. A policy that is rational when it is co-ordinated with the climate policy of other countries may be inefficient when that country decides to do things on its own.

# Rhetoric and practice

When examining the energy policy of other countries, a distinction has to be made between rhetoric and practice. Government representatives and politicians of many countries have expressed their support for far-reaching measures to reduce emissions of carbon dioxide. Following the major climate conference in Rio, many governments have put their signature to the international climate agreement. However both the statements and the agreement are vague.

Concrete measures in the form of taxes or laws are required. Statements are not enough. The only industrial countries other than Sweden that have carbon dioxide taxation are Denmark, Finland and Norway. No other countries have undertaken any significant measures to reduce carbon dioxide emissions.

So far there is a great difference between rhetoric and practice. The Netherlands for example is often seen as a pioneer in this area since they have a system of subsidies that is directed towards the reduction of carbon dioxide emissions. However the subsidies are fairly limited, amounting only to a few per cent of total energy costs. This may be compared with Denmark and Sweden where the level of energy and carbon dioxide taxes corresponds to an energy price increase of more than 100 per cent.

Several years ago, a proposal was discussed within the EU to introduce an energy tax equivalent to three öre per kg. carbon dioxide i.e. less than a tenth of the Swedish tax. The tax proposal was not approved on the grounds that it was considered to be too costly. The EU decided to await the introduction of concrete measures by other industrial countries.

In 1993 the incoming US President Clinton tried to introduce a carbon dioxide tax equivalent to 2 öre per kg. carbon dioxide. As in most other countries where such measures have been discussed, the underlying aim of the US proposal was to reduce the government's budget deficit. However the Clinton proposal failed to receive the approval of Congress on the grounds that the tax would damage US industry.

In spite of numerous political statements from a wide range of governments on the need to curtail carbon dioxide emissions, little has been done in the form of concrete measures. The countries that have introduced such measures i.e. the Nordic countries account for less than 1 per cent of global emissions. 99 per cent of global carbon dioxide emissions remain untaxed.

The taxation of carbon dioxide emissions is widely held to be an effective political instrument for the reduction of such emissions. However by acting on its own, a large part of the effect is cancelled out by the fact that the tax shifts part of the emissions abroad where it is not taxed.

Consequently an evaluation of the effects of Swedish climate policy measures must take account of the reduction in global emissions rather than simply the impact of such measures in Sweden.

143

## Import of non-taxed fuels

The import into Sweden of wood, peat, waste products and other forms of combustible material instead of oil has become a highly profitable business as a result of the taxation of carbon dioxide emissions.

Let us take a couple of examples. Pine tar pitch oil is a non-taxed by-product from the timber industry. It is often used as fuel. Since the tax on normal heating oil is nowadays very high, it has become much more economic to use pine tar pitch oil for heating purposes. Since the supply of this product is limited in Sweden, there is now a large-scale import of pitch tar oil, particularly from the USA.

Hence as a result of Swedish energy and carbon dioxide taxation, pitch tar oil is transported thousands of miles from the USA over the Atlantic to Sweden where it is used for heating. As a result of this trade, Americans have now replaced the pitch tar oil that they export to Sweden with ordinary heating oil. From a Swedish perspective, this trade gives rise to an additional cost of almost S.kr. 100 million per annum. At a global level, carbon dioxide emissions have not been reduced. On the contrary, they will have increased as a result of the emissions caused by the unnecessary transport of oil across the Atlantic. Consequently, Swedish climate policy has resulted in a substantial additional cost unaccompanied by any gain.

Another example of the effects of Swedish climate policy is provided by Swedish power stations that import olive stones from Tunisia and Spain. The effect of Swedish policy is that olive stones are transported from Tunisia and Spain at a cost of S.kr.50 million replacing coal in Swedish power stations while carbon dioxide emissions are shifted to Tunisia and Spain where coal is now burnt instead of olive stones.

There are several other examples of imports of non-taxed fuels to Sweden that are meaningless from the point of view of carbon dioxide emissions. These examples clearly illustrate the problems that arise when a country tries to take steps on its own to reduce carbon dioxide emissions. These effects are unavoidable in an open economy of the Swedish type.

## The movement of industry from Sweden

The taxation of carbon dioxide emissions in Sweden also has consequences for Swedish industry. Industrial production is on average highly energy intensive and therefore sensitive to increases in energy prices. If carbon dioxide taxes are higher in Sweden than in the rest of the world, part of Sweden's energy-intensive industry will not be able to compete with companies in other countries that are not subject to carbon dioxide taxes. Production of energy-intensive products will be moved from Sweden to other countries. Carbon dioxide emissions will be reduced in Sweden but rise proportionately in other countries. Hence as far as the effect of carbon dioxide taxes on energy-intensive industry is concerned, global emissions will be largely unaffected although Swedish emissions will be lower.

In rough terms, the "leakage" of carbon dioxide abroad is over 50 per cent. In other words, at the same time as carbon dioxide emissions have been lowered by 10 million tons per annum, they have increased by more than 5 million tons per annum in other countries as a result of the unilateral Swedish energy taxes. Accordingly, global emissions have decreased by up to 5 million tons per annum. This is equivalent to 0.015 per cent of total global emissions.

## What are the costs of energy taxation?

The next question concerns the costs borne by Sweden in order to achieve this reduction in global emissions.

The costs of carbon dioxide taxation consist partly of the costs to households and firms of using more expensive forms of fuel e.g. peat and wood instead of cheaper sources of energy such as oil and coal. Major additional costs soon arise since peat and wood is three times as expensive as coal which is the most appropriate alternative form of energy. Moreover several million Swedish crowns have been invested in adapting power stations to the need to burn wood, peat and other non-taxed sources of energy.

In aggregate, energy and carbon dioxide taxes have raised the costs of energy provision by S.kr. 6 000 million per annum as a result of replacing cheaper forms of energy with dearer ones.

The burning of wood instead of oil or coal in power stations is one of the measures that are cheapest to introduce. Here the cost is usually about 50 öre per kg. carbon dioxide. We shall use this measure as a benchmark for evaluating other types of measures.

If Swedish climate policy was restricted to the taxation of carbon dioxide emissions, costs would adjust to this level i.e. about 50 öre per kg. carbon dioxide. The great advantage of this type of tax is that firms and households acquire an interest in reducing carbon dioxide emissions in a cost efficient manner. There is an obvious advantage in reducing carbon dioxide emissions if the costs of this reduction can be kept within the range of 50 öre per kg. carbon dioxide. The government doesn't have to involve itself in the details of what households and companies should or should not do. Hence the most efficient climate policy would be some form of uniform carbon dioxide tax.

145

# Unnecessarily expensive measures

However climate policy has not limited itself to such measures. In addition to carbon dioxide taxes, the government has also undertaken measures in other areas. Investments in wood-fired power stations have for example received substantial allowances that have encouraged heavy investment in this area by central and local government. However generating electricity in wood-fired power stations is a highly expensive way to reduce carbon dioxide emissions. It often costs more than 1kr. per kg. The same also applies to the wind generators that receive government financial assistance.

An important feature of Swedish climate policy has been the use of measures directed at the reduction of carbon dioxide in road traffic. Since petrol and diesel oil are both subject to carbon dioxide taxation, it should not be necessary to take any further measures. The carbon dioxide tax automatically reduces road traffic. Moreover special measures for road traffic are particularly expensive since petrol is already so heavily taxed. It has for a long time been a major source of tax revenue for the government. To reduce carbon dioxide emissions from road traffic over and above what would be an automatic consequence of carbon dioxide taxation would cost more than 1kr. per kg. carbon dioxide

An even more expensive measure would be to replace petrol by ethanol. Once again this is an area where central and local government are investing heavily. Firstly ethanol is four times more expensive to produce than petrol. Secondly fairly large amounts of coal, oil and natural gas are used in the production of ethanol. Consequently the reduction of carbon dioxide emissions are not particularly large.

Ethanol is normally produced by means of grain. The cultivation of grain requires fertilisers. Large amounts of natural gas and coal are used in the production of fertilisers. Diesel oil is required to run the agricultural machinery and the lorries that transport the grain to the factories where it is converted into ethanol. The production of yeast and the distillation of grain into alcohol require oil, coal and electricity. In view of the considerable amounts of coal, oil and natural gas that are used in the production of ethanol, it must be a matter of some uncertainty whether or not carbon dioxide emissions will actually be lower if petrol is replaced by ethanol. An OECD report concluded that carbon dioxide emissions would be higher in the case of ethanol. A Swedish government report indicated that replacing petrol with ethanol would cost more than S.kr. 3 per kg. carbon dioxide.

Another element of climate policy is to seek to move a significant proportion of goods transport from road to rail. However it should be remembered that rail transport obtains a major part of its fuel from coal-fired power stations. If we take this factor into account, road and rail transport account for almost the same amount of carbon dioxide emissions. Granting public subsidies to rail transport in order that they can take over a growing proportion of the goods carried on the roads would be a particularly expensive method of reducing carbon dioxide emissions.

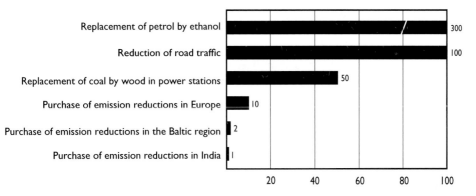

| | |
|---|---|
| Replacement of petrol by ethanol | 300 |
| Reduction of road traffic | 100 |
| Replacement of coal by wood in power stations | 50 |
| Purchase of emission reductions in Europe | 10 |
| Purchase of emission reductions in the Baltic region | 2 |
| Purchase of emission reductions in India | 1 |

20   40   60   80   100

**Figure 11.2** Swedish climate policy measures cover a wide area of costs from a few öre to 3 S.kr. per kg.

Central government and local authorities spend several million Swedish crowns each year on inefficient measures to reduce carbon dioxide emissions. It could be argued that if Sweden is to reduce its carbon dioxide emissions, it must concentrate on the measures that provide the best value for money. Some of the measures employed have tended to raise carbon dioxide emissions e.g. the replacement of petrol by ethanol.

# Why should Sweden take on a greater burden than others?

Why should Sweden as virtually the only country in the world take on an extra annual cost of several billion Swedish crowns in order to reduce global carbon dioxide emissions. If it emerges that emissions of carbon dioxide represent a major threat to the climate, the citizens of Sweden will not be more affected than the citizens of other countries. On the contrary, Sweden will be one of the countries that will be least affected.

The reduction of 0.015 per cent in global emissions brought about by Swedish climate policy is far too small to have any effect on world climate. Indeed the reduction of emissions is just too small to be measured. No one is going to thank the Swedes for this sacrifice.

## Sweden – the good example

It is often argued that a strong justification for Swedish climate policy is the effect that it would have on other nations to follow Sweden's example. Following the major climate conference in Kyoto at the end of 1997, it is clear that Sweden has failed on that point. The international agreement reached at that conference contained fairly limited and vague commitments. The idea of introducing an international tax on carbon dioxide emissions was never on the agenda. Any such suggestion had been removed long before the start of the conference. The commitments made by the participatory nations at the conference are equivalent to a carbon dioxide tax that is only a fraction of the Swedish tax.

Sweden's previous strategy to try to convince other nations to follow Sweden's example has not just been a failure but also runs the risk of becoming a costly failure. In fact Sweden may actually be forced by the Kyoto agreement to reduce its carbon dioxide emissions more than other countries.

The Swedish negotiation strategy differed from that of the majority of other countries. The latter argued that they were prepared to participate in internationally co-ordinated measures provided that a substantial number of other countries also agreed to share the costs. This should also have been Sweden's strategy.

## Benefits to Swedish industry

Similarly it is often argued that Swedish climate policy would help the country to adjust its production to meet the demands of a more carbon dioxide free economy. This would give Sweden a more advantageous economic position when other countries are compelled in a few years time to reduce their emissions of carbon dioxide.

Following the Kyoto conference, we also know that this argument in support of Swedish climate policy failed to find support from other countries. The Kyoto agreement proposed far less sweeping measures to reduce carbon dioxide emissions than have been adopted in Sweden. Consequently Swedish industry has had to adjust to a much higher carbon dioxide tax than industries in other countries will have to face in the future. As a result, the competitive position of Swedish industry has become weaker rather than stronger which was the intention of Swedish climate policy.

## Why should measures be undertaken in Sweden?

The basic problem confronting Sweden is that a reduction of carbon dioxide emissions in Sweden that would be large enough to have an impact at the global level is too expensive a measure for a small country that is responsible for only 0.3 per cent of global emissions of carbon dioxide. Not even a reduction of Swedish emissions to zero would have any effect on the global situation. On the other hand, it would seriously reduce the country's living standards.

If Sweden is prepared to invest S.kr. 6 billion in order to reduce global emissions, is it possible to find more efficient ways of doing so?

Yes, if one is prepared to pay other countries to reduce their emissions. As a result of the high levels of Swedish energy taxation, the marginal costs of reducing emissions have risen to as much as 50 öre per kg. carbon dioxide or more. Other countries that don't have high levels of carbon dioxide taxation are able to reduce their emissions at a much lower cost than in Sweden. The marginal cost of reducing $CO_2$ emissions in most EU countries for instance is less than 10 öre per kg. In East Europe, USA and many other countries throughout the world, carbon dioxide emissions can be reduced for as little as few öre per kg. carbon dioxide i.e. at about a tenth the cost of what it costs in Sweden.

Sweden doesn't need to limit itself, as is the case today, to "purchasing" reductions in carbon dioxide emissions where the price is high. Reductions of emissions may be purchased abroad at a much lower cost. For instance, Sweden has already started to reduce $CO_2$ emissions in the Baltic states on a very limited scale.

Following the Kyoto agreement, this type of measure has received international approval. Once again Swedish climate policy would appear to have failed since the emphasis has been almost entirely on the reduction of emissions in Sweden. Heavy investments have been made into converting industrial plant in Sweden in order to reduce carbon dioxide emissions. These investments have largely been a waste of money since it is possible to achieve the same effect at a much lower cost by purchasing reductions in $CO_2$ emissions abroad.

## Conclusions

Our conclusion is that Sweden has not actually had a climate policy. The entire question started off on the wrong foot since climate policy was used as a debating weapon in the nuclear power issue. It was subsequently used as a pretext for raising taxation in order to cut the fiscal deficit during the economic crisis in the early 1990s. Agricultural interests and other pressure groups have also seen their chance to use the climate issue to further their own interests.

If Sweden had taken the climate issue seriously, it would have concentrated its efforts on trying to reach an international agreement and postponed the introduction of expensive measures until the agreement had been reached. This was the policy followed by most other countries and would have saved the country's taxpayers several billions of crowns per annum on costly climate policy measures that had no effect on the reduction of global emissions of carbon dioxide.

If Sweden had wanted to demonstrate its willingness to accept the costs of reducing global emissions prior to the reaching of an international agreement, the emphasis should have been placed instead on meeting the costs of reducing emissions in other countries.

The positive lesson to be drawn from Swedish climate policy is that it provides a very clear example of the problems that arise when a small country or group of countries seeks to act on its own. Here Swedish climate policy offers numerous examples of what can go wrong.. Other countries should be able to draw important lessons from the Swedish experience.

It is also possibly the case that the various twists and turns surrounding Swedish taxation of carbon dioxide emissions may raise certain doubts in the minds of the advocates of an internationally harmonised tax as a solution to global climate issues. The Swedish experience would suggest that the introduction of a tax attracts the attention of various interest groups. Some see the opportunity to use the carbon dioxide tax revenues to strengthen the government's finances. Different groups will try to further their own particular interests group by seeking to influence the details of the tax legislation. Hyenas and vultures detect the scent of a new "tax prey". It is perhaps the global carbon dioxide tax rather than a deterioration in global climate that constitutes the real global threat.

# Commentary

# 12.

# Knowledge and controversies in the climate issue

*Bert Bolin*

The initiative by the SNS to publish this book[1] may stimulate further discussion of the apprehensions about the effects of greenhouse gas emissions on global climate. This type of initiative is invaluable but requires at the same time a balanced presentation of the issues involved.

The Intergovernmental Panel on Climate Change (IPCC) has carried out a major study of the climate issue (IPCC, 1996) which has in turn acted as the basis for intergovernmental negotiations regarding possible counter-measures. This discussion culminated in the Climate Convention meeting in Kyoto in December 1997.

Relatively few researchers have questioned the IPCC's conclusions. Their views have, however, for instance received the support of a number of industries that have an interest in the continued use of fossil fuels as well of countries that have large reserves of oil and coal. These have in turn influenced the debate on these issues, particularly in the United States. It has obviously been the wish of SNS to present the views of authors who are for one reason or another critical of the work and conclusions of the IPCC. Hence commentaries on the different presentations in this book, based on the IPCC evaluations, are essential if this book is to achieve its aim. Probably only a few readers will have ever seen the IPPC report. Even the summary (IPCC, 1996d) whose introduction is presented in Chapter 1 has probably been read by just a few.

My analysis can be mainly seen as an expression of the views that are widely held by the team of researchers that have participated in the work of the IPCC. I have also made use of more recently published work. However, the

---

[1] i.e. the originl Swedish version, from which this English version is translated.

152

conclusions drawn from this latter material should be treated with greater care since they represent the views of individual researchers and are not an integral part of a comprehensive analysis of all of the available contemporary scientific literature. Here we will have to await the publication of the IPCC's third evaluation to be available in 2000/2001.

Below, I will present my comments to the views and criticism put forward in Chapters 2-11 in conjunction with brief summaries of the major conclusions presented by the IPCC (in italics).

## The basic scientific premises

*The global mean surface temperature has increased by 0.5 – 0.6°C during this century, of which 0.3°C has occurred during the past three decades.*

This conclusion has not been challenged.

*The interaction of solar radiation and the radiation characteristics of the surface of the earth and its atmosphere determine the global climate of the earth. A number of greenhouse gases in the atmosphere of which water vapour is the most important one, has raised the global mean surface by tens of degreees C.*

**Lindzen (8)** provides an excellent survey of the greenhouse effect that is completely in accordance with the views underlying the IPCC's analysis. He is also correct when he points out that the assumption regarding the earth's albedo i.e. its capacity to reflect solar radiation plays a highly significant role when estimating the size of this increase in temperature. This explains why the values found in the scientific literature vary from 15°C to 35°C.

*Human activity has led to the emission of a number of greenhouse gases into the atmosphere. The most important of these gases is carbon dioxide whose concentration has increased by about 30 per cent since the early 1800s. The total effect of the increasing concentrations of all these greenhouse gases is equivalent so far to raising the concentration of carbon dioxide in the atmosphere by about 50 per cent. A doubling of the concentration of carbon dioxide in the atmosphere with no concurrent change in the concentrations of other greenhouse gases will raise the global surface temperature by a little more than 1°C.*

*The air temperature determines the concentration of water vapour in the atmosphere and consequently its greenhouse effect. There is therefore an important feedback mechanism whereby an increased concentration of other greenhouse gases affects the air temperature and its concentration of water vapour.*

When trying to estimate the effects of anthropogenic emissions of greenhouse gases on temperature, it is generally assumed that the relative humidity of the atmosphere remains the same which means that the concentration of water vapour in the atmosphere increases by 2 – 5 per cent for each degree

153

increase in temperature. This strengthens the greenhouse effect. In more complex climatic models, which describe the three-dimensional structure of the atmosphere as well as evaporation, cloud formation and rainfall, the model itself determines changes in the concentration of water vapour in the atmosphere. Models yield different results dependent on the assumptions regarding the physical processes having been included. However, almost without exception, they all indicate an increase in the concentration of water vapour in the atmosphere. This sensitivity of climate models to an increased concentration of carbon dioxide in the atmosphere owing to this feedback mechanism results in an increase in mean global temperatures of between 1.5°C –4.5°C (median value of 2.5°C) for a doubling of the concentration of carbon dioxide in the atmosphere. The major part of the uncertainty as indicated is attributable to the importance of water vapour in the climate system. However account is also taken here of other factors such as for example the positive feedback mechanism whereby less snow and ice enhance the absorption of incoming solar radiation. In summary: The positive feedback brought by an increase in the concentration of water vapour in the atmosphere may be expressed in simple terms as a reinforcement of the heating effect produced by carbon dioxide by 50 and 250 per cent or a median value of about 100 per cent.

**Lindzen (8)** has some reservations on this point. He has shown that this feedback mechanism is affected by the vertical distribution of water vapour in the upper part of the troposphere (3-10 km). He argues that the description of the relevant processes is too simple to allow an estimate of changes in this vertical distribution. Although Lindzen has shown the importance of accurate descriptions of the physical processes, there is no evidence either in the form of climate model simulations or direct observations to suggest that the IPCC's conclusions on the sensitivity of the climate system to greenhouse gas emissions would be affected by his arguments. The range of uncertainty given by the IPCC also includes a relatively weak feedback mechanism. There are therefore no grounds for changing the range that the IPCC has given for the sensitivity of the climate to increased concentrations of water vapour in the atmosphere. This issue will undoubtedly be discussed at length in the IPCC's third report.

*In its second report (IPCC, 1996a), the IPCC concludes that "the balance of evidence suggests a discernible human influence on the global climate".* The careful language used by the IPCC should be considered in view of the fact that it was not until recent decades that the expected increase in global temperatures due to human interference has become of about the same magnitude as that due to normal variations.

The basis for this conclusion is presented in detail by the IPCC (IPCC, 1996, p 411-443). Here the observed changes in temperature pattern up to an altitude of about 20 km and during the latter part of the twentieth century are compared with the estimated changes derived from the climatic models. This issue has been subsequently examined in greater detail by Santer and a group of scientists from six research institutes (Santer et al.,1996). This central conclusion has been the subject of lively debate during the past two years and undoubtedly played an important role during the political negotiations at the Kyoto conference.

It is important to distinguish between the carefully formulated conclusions presented by the IPCC on the one hand and the statements made by environmental organisations and politicians on the other. **Böttcher (3)** often refers to the latter but it is not the task of the IPCC to take part in that debate. On a number of occasions, I have tried to maintain this distinction when summarising the IPCC's conclusions before the scientific committee of the Climate Convention (SBSTA - Subsidiary Body for Scientific and Technical Advice) as well as at Convention conferences. It should also be noted that Böttcher quotes from a number of statements made at a climate conference at Villach in 1985 (Bolin et al., 1986). However, this conference which was largely dominated by researchers, did not propose any policy measures but instead argued the case for further research and improved co-operation between researchers and politicians in order to provide a better understanding of the problem.

It is remarkable that **Karlén (4)** does not accept the methodology that the IPCC considers to be an indispensable part of its analysis of whether or not it is possible to detect the effect of human activity on the climate. His basic premise is the following: "In order to ascertain whether the recent changes in global climate are unique and a result of human emissions of greenhouse gases … the variations in the climate over a long period of time must be known". It is, however, impossible to answer this question by only studying changes in global mean temperatures but Karlén is correct when he underlines the importance of the natural variations of the climate system.

**Karlén (4) and Ahlbäck (5)** draw attention to the finding that global mean temperatures at an altitude of between 2 km and 10 km have not risen between 1979 and 1996 while a slight increase of 0.1 °C has been recorded at the earth's surface. They conclude that the discrepancy between these different series of observations calls into question the increase in global temperatures and that human influence on the climate has been relatively limited. However, in view of the natural variations in temperature, these series of observations for less than 20 years are far too short to establish whether or not human influence has contributed to global warming. It is for example difficult to assess the extent to which observed changes have been affected by the two major volcanic eruptions in 1983 and 1991.

Ahlbeck is also incorrect when he states that the correlation estimated by Santer et al. (1996) is not significant. He also contends without any detailed analysis that the improved correlation between observed and estimated climate changes is dependent on "adaptation of model computations to data" which

155

shows that he is unaware of the ways in which new insights into important processes in the atmosphere can be integrated into a model and then tested.

As stated in the introduction to this chapter, the global temperature is obviously affected by variations in solar radiation (IPCC,1996a). However, the important question to be asked is whether or not the global warming that we have experienced during this century and particularly in the last half century is attributable to more intensive solar radiation. Lassen and Christensen (1995) have found a correlation between temperature variations and (+/- 0.2°C) and solar radiation during the last 400 years. However Laut and Gundermann (1998) have shown that the possible trend towards a warmer climate during the twentieth century has little effect on the above correlation. The conclusion that global warming during this century is due to changes in solar radiation is thus not supported. Moreover Lean et al. (1995) have found that the increase in solar radiation could perhaps explain at most about one third of the global warming that has occurred since 1960. Karlén and Ahlbeck do, however, not refute the statement that "human influence on the global climate is now discernible".

*Although climatic models still remain uncertain, they represent a synthesis of present knowledge regarding the interaction between a multiplicity of processes as well as providing a degree of internal consistency to the analysis. They are superior to the qualitative arguments that are sometimes presented in answer to the question if the successive changes in the climate are the result of human activity. A comparison with actual observations supports the view that model simulations also can be used to provide an overview of the future changes in climate that may result from different emission scenarios. Models can also be further developed as the state of knowledge about the dynamics of the climates system improves.*

**Wallin (9)** on the other hand contends that the rudimentary state of our knowledge of the role of the oceans prevents us from drawing any worthwhile conclusions about the effect of human activity on global climate change. He describes in general terms a number of important processes that are still not fully understood. This problem is widely recognised by oceanographers who have constructed the models in current use. However Wallin's criticism is far too general in nature to allow us to dismiss these models as utterly useless.

The IPCC has pointed out the need for better models. However the models in use are able to reproduce some of the most important characteristics of the oceanic circulation systems. For example the main observed features of the distributions of the carbon 14 and tritium isotopes in the sea can be deduced with the aid of available models. Leading oceanographers accept the IPCC conclusions but do not view them as final. The non-linear nature of marine systems also suggests that relatively rapid changes may occur. The importance of such phenomena are now the subject of intense discussion. This suggests, as the IPCC puts it, that "we can never exclude surprises". However this should not prevent us from utilising the knowledge that we have.

A few comments on Wallin's concluding remarks:

1. The overwhelming majority of climate researchers contend that it is likely that the earth will become warmer as a result of the emission of greenhouse gases although we do not have firm knowledge regarding the sensitivity of the climate system to continued emissions. Data from the Antarctic ice cap suggests however that the cold climate during the most recent Ice Age was partly due to a much lower concentration of carbon dioxide in the atmosphere compared to current levels.

2. According to the IPCC, climate change may be beneficial in certain parts of the world but the overall effect of these changes will be negative. A modestly rapid rise in the level of the world's oceans (half a metre to a metre over a century) would for example cause substantial damage in low lying, densely populated coastal areas.

3. The view that nothing can be done about climate change is Wallin's own standpoint. He doesn't provide any grounds for this argument. The views of the IPCC are presented in the IPCC reports (1996c, 1997)

The difference between my (IPCC) comments on the increased risk for drought over continental areas and those of Wallin is that the former are based on the analytical results of climatic models whereas the latter fails to provide any account for the basis of his argument.

*The models of the global carbon cycle that are used to estimate increases in the atmospheric concentration of carbon dioxide resulting from a given increase in emissions have been tested using a wide range of ecological and marine chemical data. The IPCC also indicates the uncertainty of the results that are provided (IPCC,1996a). However the latter is relatively limited in comparison with the uncertainty of future human emissions of greenhouse gases.*

**Ahlbeck (7)** has carried out a pure statistical analysis of the relationship between global emissions of carbon dioxide and the increase in the atmospheric concentration of carbon dioxide between 1970 and 1996. He finds that the relationship is linear. The problem is actually much more complex and must not be simplified in this manner.

The carbon cycle models that have been developed on the basis of our knowledge of physical, chemical and biological processes in the oceans indicate that the relationship between emissions and the increase in the concentration of carbon dioxide in the atmosphere is dependent on the rapidity of the increase in emissions. Until 1945, the increase was about 2 per cent per annum, between 1945 and 1970 it was approximately 6 per cent per annum while between 1970 and 1996 it had declined to around 3 per cent per annum. The emission scenario up until 2100 indicates a future increase of emissions of about 0.14 Gt per annum i.e. between 1 and 2 per cent per annum.

Ahlbecks analysis shows that the equilibrium atmospheric concentration of carbon dioxide would have been 277 ppm which accords well with observed conditions prior to the increase in emissions that began in the mid nineteenth century. However, this does not provide us with a basis for the assumption that the relationship between emissions and atmospheric concentrations can also be used to estimate future concentrations. He also points out that the results are sensitive for the higher order terms. However these cannot be determined with data that covers only 25 years. They would anyhow not be applicable for estimates of future atmospheric concentrations.

The estimated atmospheric concentration of $CO_2$ provided by the IPCC for the year 2100 is about 700 ppmv (under the assumption that emissions have increased to about 20 GtC per annum) is the best estimate that we have today but is obviously still uncertain. However higher figures are just as probable as lower ones. Ahlbecks estimates which are much lower are not credible.


## Economic issues

*We need images, scenarios of probable future emissions of greenhouse gases in order to estimate possible climate changes i.e. scenarios of future world economic development. It is obviously not possible to predict this development. Consequently, the IPCC (1992) has constructed six basic scenarios offering a wide range of assumptions regarding population growth, global growth of GDP, the efficiency of energy use, availability of fossil fuel ( regarding both reserves and resources) and the costs of using alternative sources of energy such as hydro electricity, wind power, bio-energy and solar energy. The various future scenarios presented in this manner obviously implies increased uncertainty regarding the climate changes that can be expected if no counter measures are adopted. The IPCC has conducted a thorough analysis of the significance of this uncertainty.*

**Gerholm's (6)** main thesis is that due to uncertainty about the future, we don't know anything about the climate that we will have hundreds of years from now. Therefore we ought to wait for at least ten years before mitigating a possible future climate change. It is the view of the IPCC that we know that the world will probably become warmer. If the climate system is relatively insensitive, it may take some considerable time before global warming becomes a threat. However the changes may be substantial and come about very rapidly. This would suggest that considerable caution will have to be employed initially.

Gerholm is critical of the IPCC scenarios, perhaps due to the mistaken impression that meteorologists have been responsible for the IPCC assessment studies. This is incorrect. The basis for the 1992 scenarios was limited but the analysis carried out in 1995 of the socio-economic issues involved (IPCC 1996c) cover 450 pages and include the work of some of the leading experts in the field. The IPCC scenarios are open to criticism. A decision was, however, made already in 1996 that a set of new scenarios be prepared. The Energy Forum at Stanford University (USA) plays an important role in this work.

The assumptions used by the IPCC regarding the total supply of fossil fuels are approximately the same as those used by the WEC (World Energy Council) and by Gerholm although the latter ignores that there are much larger resources available than those that are actually known (IPCC 1996c, p87.) It is probable that they will be exploited during the next century when the population of developing countries is expected to be twice today's levels.

The IPCC's presentations have not been misleading as Gerholm means. However, his analysis of the consequences of annual emissions totalling 14.4 billion tons carbon by 2100 compared to the IPCC's central figure of 20.3 billion tons suggests that he doesn't live as he learns. His scenario falls within the IPCC's range of uncertainty. If different estimates for the sensitivity of the climate system are considered, we obtain two sets of scenarios for the year 2100, showing a change in temperature amounting to 1.0 –2.9°C and 1.2 –3.5°C respectively i.e a reduction of 0.2- 0.6°C. In other words, this reduction is small compared with the uncertainty surrounding the extent of the change in temperature.

*The IPCC's conclusions imply that we currently see perhaps less than half of the changes for which emissions of greenhouse gases will be ultimately responsible. This is partly due to the inherent inertia of the climate system and to the fact that warming is offset by aerosols created by the sulphur emissions from burning coal and oil. However, when the carbon dioxide concentration in the atmosphere is stabilised, with or without efforts to mitigate, the aerosols disappear, while the greenhouse gases (except methane) remain for more than one hundred years, due to the climate system's slow rate of adjustment.*

*The introduction of counter measures at an early date will create increased flexibility. The IPCC argues that "the challenge confronting us now is not to find the best energy policy for the next hundred years but to choose a sensible strategy and then to gradually adjust in the light of new knowledge". Regarding the choice of counter measures, it is a matter of balancing the economic risk of undertaking measures at too early a stage e.g in the form of capital losses, against the risk of postponing the measures which may ultimately lead to more costly measures being required rapidly at a later date.*

*The possibility of using traditional cost benefit analysis as an aid to decision-making is discussed at length by the IPCC. The difficulties are substantial due to the uncertainty regarding the costs of the measures and the value of the reduction of future damages. Although the former may be assessed with some degree of certainty today, it is much more difficult to predict their changes in the future as a result of technical development. It is an even harder task to estimate the costs of damage at a future date. Moreover there is no agreed method of how to assess the value of ecological or cultural changes, nor the effects on human health. There is a sizeable risk that the costs of counter measures will be overestimated while the costs of the damage are underestimated.*

**Gerholm's (6)** contention that it doesn't really matter if we wait ten years

159

before taking a decision is misleading. He fails to take into account the full consequences of waiting to introduce a package of counter-measures. The level of emissions in ten years time will be higher than would have been the case if measures were undertaken earlier. Assuming that the use of fossil fuels is not reduced at a more rapid rate, this will in turn lead to a higher level of emissions far into the next century.

**Radetzki (10)** discusses the climate issue entirely from the perspective of cost-benefit analysis. This is a valuable contribution to the debate and provides a more concise description of the issues involved. It is based on the IPCC evaluations and illustrates the difficulties in reaching meaningful results. It also helps to explain the doubts that the IPCC have had in relation to assessments of the need for measures on the basis of cost-benefit analyses. The costs for mitigating a doubling of the concentration of carbon dioxide in the atmosphere is estimated to be equivalent to 4 – 9 per cent of GDP in the developing countries and to 1.5 per cent in the advanced industrial countries. However the estimated global average of approximately 2.5 per cent tends to underestimate the impact on small economies whose "weight" is limited in international terms. Moreover it is open to question whether a method of estimation that has been developed for the OECD countries can be simply applied to a completely different set of countries in the developing world. Moreover it is hardly meaningful to project these estimates far into the future. Over a century, a multitude of positive and negative feedback mechanisms can arise for which we have no current knowledge.

The IPCC has shown that there are many opportunities to reduce energy use and thereby emissions of carbon dioxide. The begin with, the costs are small and sometimes such measures may even be profitable (IPCC, 1997). This is noted by Radetzki when he refers to the work of Repetto and Austin (1997). A surplus would require above all:

1) an increase in flexibility throughout the economy that would reduce the costs for adjustments,

2) a reform of the tax system and/or the introduction of trade with emission rights, and

3) a focus on measures that are also of value for other reasons (for example, the reduction of air pollution).

The latter is what is usually called "win-win situations". In this context, it is interesting to note that in the 1990s, Germany reduced its carbon dioxide emissions by improving the efficiency of east German industry, while the UK replaced coal with natural gas. The lowest per capita emissions in the EU are in France due to the predominant role played by nuclear energy in the production of electricity. It is remarkable that Redetzki dismisses such opportunities, stating that "there are no grounds for the optimists' views that climate policy intervention is without costs". It is, however, important to make use of all available opportunities, but in the longer run measures to reduce emissions will not be free of charge. alternative sources of energy are still generally more expensive than fossil fuels, although technical innovation continues to lower the costs for their use.

Most economists assume that in the foreseeable future, the industrial countries will have an annual growth rate of about 2 per cent or above. This will mean that GDP would double by the middle of the next century. There is undoubtedly a lot to be done, not least to improve welfare and create a more egalitarian society, tasks that will require increased economic resources. It would be remarkable if not a relatively small proportion of these increasing resources, perhaps a few per cent, could be set aside for dealing with global environmental problems. The question should rather be *how* than *if* such measures can be undertaken.

**Gerholm's** final conclusions are in many respects in agreement with the IPCC's latest report. But in the absence of policy intervention, low prices for oil, gas and coal will prevent the exploitation of new sources of energy. Given the economic premises, the market will adopt a short-term perspective on this issue. Sweden should hardly operate on its own but rather in close co-operation with the EU and globally within the terms given by the Climate convention. From experience we know that political negotiations are time-consuming. Therefore it is important that measures be undertaken sooner rather than later. Policies that are well formulated need not initially give rise to substantial costs.

**Edin (11)** argues with great conviction that Sweden should not develop its own climate policy. The policy that has been followed during the past decade has undeniably not always been clear and administrative regulations have dominated. The IPCC has contended that economic policy instruments are superior to administrative controls, although it seems likely that a mix of the two approaches will be preferable.

A much more open attitude is required regarding the international situation and the seriousness of the problem than is shown by Edin. In addition to the need to create forms of co-operation between the advanced industrial countries and the less developed world, which undoubtedly is the single most important problem in the international arena, the development of methods of how to use economic policy measures in a global context has become an issue of high priority. It was at the top of the agenda at the last meeting of the Convention in Buenos Aires in 1998. How should then Sweden react?

I also take strong exception to Edin's view that it doesn't really matter what Sweden does since her emissions of greenhouse gases are of little consequence in an international perspective. As long as out emissions are higher in per capita terms than on average those of developed countries (at present, three times as high), we have a responsibility in terms of international solidarity to reduce them. The climate issue cannot be solved without some measure of solidarity between nations.

161

# Science and politics – the credibility of the IPCC

The IPCC's role in the evaluation work that has formed the basis of the creation of the Climate Convention as well as its later work is the subject for some discussions.

*The IPCC is an intergovernmental panel founded by the United Nations Environmental Programme (UNEP) and the World Meteorological Organisation (WMO). Its responsibilities have been laid down by the parent bodies. The IPCC is independent from the Climate Convention, although it has co-operated with the Convention on scientific and technical issues, since its creation. The IPCC's work is financed by a trust fund that receives voluntary contributions from interested countries, almost entirely from the industrialised world. The secretariats of the three working groups have been financed by the respective host countries, between 1992 and 1997, the UK, USA and Canada.*

*The IPCC has well-developed procedural rules for its work (a summary of these is available in IPCC 1997, inside page). The national delegates to the IPCC decide on the overall structure of IPCC reports and the work is then carried out by three working groups. The authors of the individual chapters in the report – there were almost 300 authors involved in writing the 49 chapters of the 1995 report – are chosen by the working group's executive committees on the recommendation of participating countries. Each team of authors should contain at least one researcher from a developing country. The IPCC reports are only based on published reports and articles.*

*The draft versions of the different chapters are reviewed, in a first round by a team of experts who were not involved in the writing of the report and then by another group of researchers selected as representatives for their respective countries in the particular scientific-technical areas concerned. The authors then have the responsibility for taking into account the various viewpoints that have been expressed, provided that they are scientifically viable. Controversial issues should be considered explicitly. The Summary for Policy Makers is approved at an IPCC committee meeting where researchers have the right to reject proposals that are not scientifically well founded. Moreover, the main report remains the responsibility of the authors to the individual chapters.*

*In my view the IPCC has worked well and the researchers who have participated in this work have done so with enthusiasm and commitment. a number of the authors have devoted several months work to this task. There has naturally been an in-built tension in this process between on the one hand, the desire of researchers to carry out objective research and on the other, the risk that national interests would influence conclusions. This becomes particularly evident in the final discussion of the summaries. As chairman, it has been my primary duty to argue in support of the integrity of researchers, to minimise conflicts and to try to establish an atmosphere of mutual trust between the research community at large and the IPCC representatives of individual countries. It is up to others to judge whether or not this work has been successful.*

162

Although **Moberg**'s (2) study of the interaction between science and politics is general in nature, it raises a number of interesting principle points. The possible relationships that he describes emphasise the importance of viewing the climate issue in a wider social and political perspective than is usually the case.

In his closing sentence: "If these assessments are correct, it would also be reasonable to assume that the environmental threat is exaggerated and that substantial resources are being unnecessarily invested to try to avert it." This question has been examined at length by the IPCC (1996c). Moberg points to the difficulties of reaching agreement on a common approach to public goods and that measures may be insufficient even when the threat of a climate change is perceived as real. Accordingly his contribution should be seen as an appeal for awareness of the multiplicity of factors that exert an influence on the way in which individuals, groups and countries act when faced with the threat of climate change.

Moberg further states that "with exception of the statement that the threat is serious, no value judgements are involved". In reality, the climate issue is characterised by a number of value judgements that remain influenced by the country's geographical position, level of development and not least by the way in which individuals experience this threat.

Prestige, status and economic interests certainly exert an influence on the behaviour of most people. Researchers – and the authors of this book – are certainly not oblivious to these considerations. However, Moberg presents no evidence for his statement that "researchers are largely in the hands of the government". The leading researchers, for example at Princetown University in the USA, the Max Planck Institute in Hamburg and the research departments of the Meteorological Office in England where the most advanced work is carried out in climate modelling, act as researchers, otherwise they would soon lose their credibility. Moberg is obviously unaware of the openness that is demanded by the research community when presenting research in academic journals and at conferences, a factor that substantially reduces the risk of bias.

**Böttcher**'s comments on these issues are largely based on the work of Boehmer-Christiansen (1997 and earlier). He has not carried out any detailed research himself into the social and political role of the IPCC. Relatively few researchers have analysed the climate issue from the perspective of social science. Still, Boehmer-Christiansen's is just one, and several other studies are underway. I am sceptical of many of her statements and accordingly to those of Böttcher as well. Boehmer-Christiansen contends that researchers mainly pursue their research because of money, power and status. Her aim is "to show that the IPCC should be studied as a political institution that acts to defend the interests of scientific and research organisations rather than to protect the environment". This statement remains, however, as a largely unsubstantiated hypothesis. The formulation of this hypothesis also suggests a bias in a her

163

approach. She does not have a firm understanding of the scientific issues and has no evidence to support her contention that the work of the leading research institutions is primarily oriented towards the fulfilment of political goals. In her view, researchers who work for the IPCC are controlled by organisations such as the ICSU, WMO and the UNEP. This is quite absurd to anyone who has any knowledge of how scientific work is pursued.

A few additional comments may be of interest in this context. Böttcher states in his introduction that the dangers of a colder climate as a result of an increase in atmospheric pollution were high on the scientific agenda in the early 1970s. This is hardly borne out by a study of the academic literature. The importance of increasing levels of greenhouse gas emissions in the atmosphere was then already at the centre of interest. This is evident from the conference that was organised at that time as part of the Global Atmospheric Research Programme (GARP, 1974). A good portion of the leading scientists in the field of climate research took part in this conference.

Böttcher also quotes a number of statements made at a conference in Villach in 1985. It is, however, evident from these quotations, as indeed from the conference summary, that this conference, being dominated by researchers, did not recommend policy actions but rather further research and an improved co-operation between researchers and politicians in order to understand the new issue that was emerging. The need for a better assessment of the state of scientific knowledge in this area was one of the main reasons for my own participation in the work of the IPCC. In my view, it is of fundamental importance that scientists do not behave as activists. for example, that argument that the hot, dry summer in the USA in 1988 was caused by an enhanced greenhouse effect, as was maintained by some researchers a few years later, was not adequately supported by scientific evidence.

However, it is evident that Böttcher has not read the IPCC reports when he states that the IPCC means "more of the same thing". His own loosely based statements are not supported by the scientific studies on which the IPCC's conclusions were based.

Böttcher's accusation that "scientific agreement (within the IPCC) is achieved by voting" is similarly unsupported by any evidence. Uncertainty is reflected in the form of the confidence intervals that are presented. Nevertheless, I do agree with the argument put forward by Hansson and Johanneson (1997) and quoted by Böttcher, that divergent opinions should be presented with greater care. The new chairman of the IPCC also shares my opinion. I do, however, not agree with the view, as put forward by Böttcher that the IPCC has introduced bias into the debate. Few, if any, academic studies support this contention.

Neither Moberg's nor Böttcher's chapter provide any support for the criticism of the IPCC's evaluation that has been put forward.

# What were the results of the Kyoto negotiations?

The negotiations in Kyoto on a supplementary protocol to the Climate Convention were in some respects successful. However with regards to reductions in greenhouse gas emissions, little progress was made (Bolin 1998). The IPCC assessment remained as the basis for the scientific judgements although scientific issues were hardly discussed by the national delegates to the session. Political and technical issues were rather predominant. The delegates of the various parties sought to secure negotiating positions that would minimise disadvantages for their own countries. The threat of a climate change was recognised by almost all countries but was nevertheless not considered to be sufficiently threatening to justify the introduction of more far-reaching measures to reduce emissions in the short term. Politicians obviously fail as yet to appreciate the serious implications of the inertia of the climatic and social systems even though they are clearly illustrated in the IPCC reports.

The EU, USA and Japan have committed themselves to a reduction of greenhouse gas emissions by 2010 of 8, 7 and 6 per cent respectively, compared to 1990 levels. For the United States, this represents a reduction of present levels of emissions of between 15 and 20 per cent. At the same time Russia and the Ukraine were allowed to return to the emission levels in 1990 which represents a 30 per cent increase compared to present levels. It should also be noted that industrial countries have reduced their emission levels by 5 percent between 1990 and 1995, primarily as a result of this reduction in eastern Europe. This also became the new target for 2010. This agreement therefore mainly implies a redistribution of emissions between industrial countries.

Industrial countries also agreed to the international trade of emission rights. A redistribution of emissions will presumably be based on this trade which will create decided advantages for the former communist countries of eastern Europe, particularly Russia and the Ukraine. Developing countries, on the other hand, are still permitted to increase their emissions of greenhouse gases until 2010 by which time their total emissions may well have reached the total emissions of the industrial countries. However, in per capita terms their levels of emissions will still be about one quarter of those of advanced industrial countries, which is mainly due to their continued rapid increase in population, although this rise is now tending to decline.

As a result of these limited commitments, the concentration of carbon dioxide in the atmosphere will continue to increase unabated during the first decade of the twenty first century to a level of about 380 million parts per million i.e. to more than 35 per cent above the pre-industrial level. It will be interesting to see whether any marked signs of environmental damage from climate change will be possible to identify during this period and whether negotiations on radical counter-measures will be required.

# References

Boehmer-Christiansen, S.(1997), Uncertainty in the Service of Science: Between Science Policy and the Politics of Power. I G. Fermann (ed.) *International Politics of Climate Change*, 53-82. Scandinavian University Press, Oslo

Bolin, B. (1998), The Kyoto Negotiations on Climate Change: A Science Perspective. *Science*, 279: 330-331.

Bolin, B., Döös, B., Jäger, J. och Warwick, R. (1986), *The Greenhouse Effect, Climate Change and Ecosystems*. 541 sid., John Wily & Sons, Chichester.

GARP, 1974. *The Physical Basis of Climate and Climate Modelling*. GARP Publications series, No 16. ICSU and WMO, Geneva.

IPCC (1996), *Climate Change 1995*, Cambridge University Press

IPCC (1996a), *Climate Change 1995 - The Science of Climate Change*, Cambridge University Press

IPCC (1996b), *Climate Change 1995 - Impact, Adaptation and Mitigation of Climate Change*, Cambridge University Press

IPCC (1996c), *Climate Change 1995 - Economic and Social Dimensions of Climate Change*, Cambridge University Press

IPCC (1996d), *IPCC Second Assessment - Climate Change 1995*. IPCC, Genève.

IPCC (1997), *Technologies, Policies and Measures for Mitigating Climate Change*. Technical Paper I. World Meteorological Organisation, Genève.

Lassen, K, Friis-Christersen, E. (1995). Variability of the solar cycle length during the past five centuries and the apparent association with terrestrial climate. *J. Atm. and Terr Phys.*,57: 835-845.

Laut, P., Gundermann, J. (1998), Does the correlation between solar cycle lengths and Northern Hemisphere land temperatures rule out any significant global warming from greenhouse gases? *J. Atm. and Terr. Phys.*60: 1-3.

Lean, J., Beer, J., Bradley, R. (1995), Reconstruction of solar irradiance since 1610: Implications for climate change. *Geoph. Res. Letters*, 22: 3195-3198.

Repetto, R. och Austin, D. (1997), *The Costs of Climate Protection: A Guide for the Perplexed*. World Resources Institute, Washington DC.

Santer, et al. (1996). A search for human influence on the thermal structure of the atmosphere. *Nature*, 282, July 4.

# Biographical
# Notes

# Biographical notes

*Jarl Ahlbäck*, has been Associate Professor in Environmental Engineering at Åbo Akademi University since 1993. He was a member of the Finnish parliament's committee of experts, the Energy Policy Council, between 1983- 1988 and chairman of the Council's Energy Savings Committee. He was awarded his doctorate in technology in 1989. Ahlbäck has published a number of articles on the purification of carbon dioxide emissions and process design with the aid of statistical methods. He lectures in environmental technology at Åbo Academy University and is a monthly columnist on the Finnish daily newspaper, "Åbo Underrättelser". He has also taken part in TV and press debates on the greenhouse effect.

*Bert Bolin*, Professor of Meteorology at Stockholm University between 1961- 1990, has been involved in the development of the Global Atmospheric Research Programme since 1965 (predecessor of the World Climate Research Programme) and the International Geosphere Biosphere Programme from 1985. He was Chairman of the Intergovernmental Panel on Climate Change, the IPCC from its formation in 1988 until 1997. Bolin has been an advisor to the Swedish government on research and environmental issues since 1983, received the WMO Prize in 1981, the Tyler Prize in 1988 and the Blue Planet Prize (Japan) in 1995. He is a member of the Swedish Academy of Sciences and the Swedish Academy of Engineering Sciences as well as the US National Academy of Sciences.

*Frits Böttcher*, Professor of Physical Chemistry at the University of Leiden, 1947-1980, is chairman of the board of the Global Institute for the Study of Natural Resources in the Netherlands. He was the first chairman of the Netherlands Science Policy Council between 1966-1974 and held leading positions in several of the country's science councils. He was chairman or a member of the board of several companies and one of the founders of the Rome Club founded in 1968 of which he became an honorary member in 1986. He has published over one hundred articles in scientific journals and has lectured in Europe, USA and Japan.

*Karl-Axel Edin*, Ph.D. in theoretical physics, works as a consultant in his company, Tentum Energy Ltd. He was head of research at the former National Board of Energy and subsequently Chief Executive of Kraftsam, an organisation representing the interests of the energy industry. He has taken part in several committees of inquiry into energy policy between the mid-1970s and

168